D0149387

THE POWER OF
REST

Why Sleep Alone Is Not Enough:

A 30-Day Plan to Reset Your Body

MATTHEW EDLUND, M.D.

HarperOne
An Imprint of HarperCollinsPublishers

HarperOne

THE POWER OF REST: Why Sleep Alone Is Not Enough: A 30-Day Plan to Reset Your Body. Copyright © 2010 by Matthew Edlund, M.D., M.O.H. All rights reserved. Printed in the United States of America. No part of this book may be used or reproduced in any manner whatsoever without written permission except in the case of brief quotations embodied in critical articles and reviews. For information address HarperCollins Publishers, 10 East 53rd Street, New York, NY 10022.

HarperCollins books may be purchased for educational, business, or sales promotional use. For information please write: Special Markets Department, HarperCollins Publishers, 10 East 53rd Street, New York, NY 10022.

HarperCollins Web site: http://www.harpercollins.com
HarperCollins®, 🏭 ®, and HarperOne™ are
trademarks of HarperCollins Publishers

FIRST EDITION

Library of Congress Cataloging-in-Publication Data is available upon request.

ISBN 978–0–06–186276–2

10   11   12   13   14   RRD(H)   10   9   8   7   6   5   4   3   2   1

CONTENTS

Part 1
THE 30-DAY PLAN
TO RESET BODY AND MIND

Part 2
PUTTING IT TOGETHER— MAKING LIFE MUSICAL

part 1

THE 30-DAY PLAN TO RESET BODY AND MIND

chapter 1

WHY YOU NEED ACTIVE REST

I used to think rest was a waste of time. I'm a doctor—why rest when you can be seeing patients, teaching, and writing research papers? It took me a long time to figure out that most medical people and far more of the general public have gotten the basic facts wrong. Rest is not a waste of time. It is a biological need—a process for restoration and rebuilding. Rest is not useless but a major pathway to our renewal, our survival. The process of aging need not mean the inevitable decline of a youthful healthy human machine into a rusting, decayed hulk. Rather, we should see aging as a script in which the body experiences, learns, rebuilds, and regrows (including our brains) throughout our lives. We re-create, renew, and reorganize ourselves through the process of rest.

Why is the importance of rest so ignored? Partly because we spend so much time working with, for, and around machines that we start to think we are one. We're not. We're living beings. In the shortest moment we can do more than any machine and certainly enjoy ourselves a lot more. One of the basic design elements of the human body is its requirement for renewal through rest. We need rest to live, just like we need food. Once we understand even a little of how our body is designed we can get far more done in less time, achieve wildly different peak experiences, and make our sometimes scattered, too often exhausted lives rhythmic and musical. If we know how to rest, the simplest acts can become moments of pure pleasure. Rest is primarily an active process that makes us vital and (re)creative. No, you don't have to feel tired all the time. You just need to know how your body is designed. Use your body the way it's designed, and you can live happily, long, and well.

But I'm getting ahead of myself. It took me many years before I understood how important rest is to human design, health, and pleasure. For a lot of those years I thought the big problem was lack of sleep.

San Diego

Decades ago I was standing in a lunch line at the San Diego VA Hospital. It was well past 1:00 p.m., and I was very hungry. I was an internal medicine intern, regarded by those above me as a lower life form able to perform and learn without much sleep or other forms of rest. Working one hundred ten hours a week was routine. I had taken six admissions after midnight while dealing with the frequent emergencies of our regular patients. Work had started at 7:00 a.m. the previous day, and when the last admissions rolled in after 6:00 the next morning one of us, Dr. Harvey Motulsky, said to me, "Another opportunity to excel," and smilingly tottered over to see what new clinical disaster awaited him. Over the previous twenty-four hours none of us had slept more than a few minutes.

I comforted myself that the yellow ticket I gripped in my hand meant I didn't have to pay for the meal and was eyeing a choice of three potential entrees (what in the world could be under that breading?) when the cafeteria's loudspeakers blasted a code blue.

Someone was trying to die. Our job was to stop them. I had to go. For a moment I wondered how much food I could stuff down within thirty seconds, but then I realized I'd waste time going through the checkout line, and I needed to go *now.*

Another code blue crackled off the loudspeaker. Now two people were trying to die. Which floor should I go to? I dumped my tray and ran.

That's what we did in those days. We ran. We ran day and night. It was thirty-six hours on call, twelve off. The next day we got in and worked twelve to fifteen hours, got some sleep, and then went back for thirty-six hours more, seven days a week. Many interns would finish work, go to the parking lot, and stare at the building all confused. They could not remember where they put their cars. Driving home was not easy. One evening I was stopped by the San Diego police on Interstate 5. They told me I was weaving and thought I might be drunk. Then they saw I was wearing hospital whites and let me go.

Though the standard workweek was one hundred ten hours, our professors told us we were very, very fortunate. We were on call every *third* night. In their day the rule was to be on call every *other* night. Thirty-six hours of work, twelve off, then repeat. The standard joke was that being on call every other night meant you missed half of the interesting cases.

The message we learned was that sleep was for sissies. You had to be ready for anything, any emergency or disaster, twenty-four hours a day for the rest of your life. And you could never, ever make a mistake If you made a mistake, someone might die.

Lots of people did die. In this country alone tens of thousands have died from unnecessary mistakes. They died in large part because interns and residents were so sleep deprived they screwed up magnificently, making errors like alcohol-fueled skiers trying to race down a mountainside.

Laws now cap trainee work schedules at eighty-four hours a week, but the medical mind-set has not really changed. The laws are often circumvented, and doctors in training continue to make uncountable clinical errors. Many medical teachers, including medical school deans, still think sleep is a waste of time. At the 2009 Associated Professional Sleep Societies' meeting in Seattle, one of the world's best known sleep researchers explained that he had told a medical school administrator that the reason a medical student kept falling asleep in class was because he had narcolepsy, a rare condition in which sleep uncontrollably overpowers you. The administrator did not ask how to treat or accommodate this prized pupil. He wanted to know how to get rid of him.

Work can become destiny. I trained in internal medicine, then public health, then psychiatry before I found out that sleep medicine was a real and possibly viable profession. I became a sleep doctor and ran a division at Brown Medical School before setting up a sleep lab in Florida, while managing to make my living as clinical director and then medical director of a hospital. I taught others how biological rhythms can be used to improve health and performance, and I helped people get well.

But improving sleep was not sufficient. Many millions sleep well and still feel tired, exhausted, and frustrated all day long—for much of their lives. They don't know what it is to feel fully alert, awake, and alive, excited at what will happen the next moment. In order to fully know the peak experiences people describe as their best memories, indeed just to heal and survive illness, people need to learn how to rest. I discovered that fact after taking care of thousands of patients. One of them was Kelly.

Kelly

She walked into my office slowly and sat down painfully. Her major complaint was sleeplessness. Kelly could not sleep night or day. She was always exhausted, always fatigued. Beyond slumping into a chair, there was precious little Kelly could do.

I asked her what had brought her to see me. Kelly's story had an uncomfortably familiar ring.

She had been moving up the ladder in a huge quasi-governmental corporation famous for driving its employees insane. Jealous coworkers told her to her face that promotions arrived like clockwork because she slept with the bosses. The slurs made her work harder.

Most workweeks she put in at least sixty hours. If there were crises in the middle of the night or during a holiday, Kelly drove in and fixed them. She loved solving problems. Once she had been very good at it.

When she returned home from work each day she continued the "crazy busy" pace, socializing with her husband and friends, decorating and remodeling the house, studying books and journals so she might do her job better. The more she worked, the more her responsibilities grew. Kelly loved the challenges.

The first major problem she noticed in herself was lack of mental concentration. Soon she was also unable to sleep. She woke up all through the night, stared at the clock, and tried to go back to sleep. If sleep returned, it was fitful. She would try to drag herself out of bed, forcing her body to work, until she could not get up.

By the time she arrived in my office Kelly was an emotional and physical wreck. She had not been able to work for over a year and was living on medical disability. Her diagnosis was chronic fatigue, a syndrome many physicians refuse to acknowledge exists. For years she moved with increasing frustration from doctor to doctor. She had not lost her marriage but had lost many friends, plus the ability to do nearly anything that gave her pleasure.

Kelly looked at me, then down at the floor. She began to cry. She was young but didn't look it, her face pale and pasty, her arms bloated and matte white.

I asked Kelly about her day—when she woke, what she did, when she tried to sleep. Shifting her medications helped. Changing sleep behaviors helped. She began to sleep and said she slept well.

I had been taught and believed that sleeping well gives people more energy to do what they like. That should have been true for Kelly, but it wasn't. She remained tired and fatigued most of the time. Perhaps she was like some of my other chronic fatigue patients, who if they do too much pay for it later with paralyzing exhaustion. Yet even after sleeping well, Kelly often could not get up from the bed.

Others had told me that my therapies had improved their sleep mightily but that they still felt tired and exhausted all day. Actually, there were a lot of people who said that they now slept well but they still felt dull-witted, slow, and tired. Most of those patients did not suffer from chronic fatigue. Clearly, sleeping well was not enough to make them function effectively.

I needed something else, something that would help people function better throughout the night and day. I taught Kelly about a revolutionary way to structure her daily life—going FAR. Going FAR means using food, activity, and rest in a sequence that is repeated throughout the day. There are many advantages to going FAR, which works because the human body operates through basic, underlying rhythms. Ever wonder why people like music? It may be because each of us is fundamentally musical, because rhythm is one way our body's cells communicate with each other. One benefit of going FAR is that it is a first step toward making your life musical. You'll learn about that in this book.

As I learned to treat people by going FAR, I also began to recognize the dual nature of rest. There are passive forms of rest, like sitting in front of a TV. But sitting or lying in front of a TV is not really restful or restorative for many of us. The full powers of rest as restoration come through the different forms of active rest. Active rest consists of directed restorative activities that rebuild and rewire body and mind. Active rest lets you retune and reset, consciously directing your body and brain to become more capable of doing whatever you want to do. As I discovered new active rest techniques, I found that active rest could make people amused in the middle of the most boring days as well as give them new tools to attain peak experiences. You'll also learn many of these active rest techniques in this book.

Today Kelly is happy, quick, almost always sharp. Sadly, she can't go back to her old job, as her chronic fatigue exhausts her by late afternoon. But she can work. She now helps run a company she formed with her brother, and she writes articles and books. She spends most of her time with people she cares about. She's learned how to pace life, following the rhythms of her own powerful biological clocks that set the basic conditions of every cell in her body. When she wakes each morning, she looks forward to the day. She knows that rest is required for healing, for rebuilding, for health and life itself.

Too many never get to enjoy life this way. Too many people from their adolescent years on feel tired, fatigued, or exhausted day after day. I expect my sleep patients to complain about fatigue but unfortunately, they're not the only ones complaining. So do most of my friends. Tiredness and fatigue are now common side effects of American life.

It doesn't have to be this way. Not at all.

What works for Kelly will work for almost everyone. Learn to rest using the inner wisdom of the body, and what had been impossible becomes possible. The human body usually knows what to do to keep us healthy. It takes years to completely reproduce heart cells, but on the subcellular level recent research shows we essentially rebuild the heart in three days. Many critical heart proteins last only thirty minutes before they are recycled and their parts reused. If you give your body the right conditions and the tools it needs, especially the benefits of rest, it can rebuild and renew itself. Debilitating illnesses like exhaustion syndrome, chronic fatigue, and fibromyalgia should be rare. In many cases they can be prevented if you know how to rest well. But rest can do far more than prevent illness. Resting right can help you simultaneously feel fully alert and fully relaxed. When you learn to do the quick, easy rest techniques described in this book, then brief, random moments may help you to create a joyous, comforting life.

Why Rest Is Much More Than Sleep

Ask yourself this: if you could do less and become healthier, more productive, and successful, would you do it? You can, if you know how to rest.

Most people I know have curious ideas about rest. When I ask patients, colleagues, and friends what they do when they rest, they mention two things: (1) sleep and (2) watch TV.

Sleep used to take up over a third of human life, and the U.S. Census Bureau reports Americans spend half their leisure time sitting in front of the television set. Yet sleep and TV are just two types of *passive* rest.

The truth is that rest is restoration. Your body is not some machine that hits its peak at twenty and slowly (okay, sometimes quickly) burns out and falls apart. You are a living organism. You remake yourself every day. You are changing with every moment, learning through doing, and your body needs to rebuild and renew, keeping intact those new lessons you receive with each day of life. To live is to change, and you change through rebuilding. The way you do a math problem at age sixty-five is totally different from how you did it at twenty-five, but you may do it faster and more accurately because you now have a brain that has learned, experienced, and rebuilt itself through the years. You do lots of that inner rebuilding when you rest.

Looked in the mirror recently? For a moment, take a close look at your skin. Remember that well-regarded face.

Wait two weeks. Look again.

Chances are really good that you still look like you. You might even think nothing has changed.

Except it has. Almost all of your skin is brand-new. The skin of your face has been entirely *replaced*. That's what rest does. During rest the body rebuilds, renews, rewires, and re-creates itself.

Your skin rebuilt itself during the process of passive rest. If you were going FAR and got enough sleep and the right foods at the right times, the process of skin renewal went on very well. But it happened unconsciously. You didn't have to think about it. Active rest requires thought and direction. Fortunately, active and passive rest may coexist throughout the day. Even better, they can aid each other enormously. Let me illustrate.

To engage in social rest, one of the four main forms of active rest, you invite a colleague to walk with you to lunch (social rest technique 4, discussed in chapter 5). As you stroll by a few stores, checking out what's new and what's on sale, you round a rough corner and slip as you cross the street. Your colleague comments that your mutual boss has recently been looking really tired. Perhaps all the administrative changes brought by the latest economic crisis are keeping him up at night. Then she tells you about her spouse and kids, and mentions the annoying noise made by trucks hauling building materials to a new school down the block from her house. You pass two restaurants you know all too well, check the specials, then move on to the third, where the outdoor blackboard announces a grilled salmon you want to try. There are no empty tables outside, but as you walk inside you request a spot by the window.

You sit.

As you sit down, intracellular messages race through the skin cells of your face and hands. Though you wore long sleeves and pants, the thin layer of sunscreen you put on that morning was not deep enough. The midday sun is strong, and ultraviolet rays have baked your cells, destroying outer layers and slicing into their DNA. Special repair enzymes push out from chromosomes to fix the DNA damage while other proteins move forward to replace broken internal cell membranes. If the repair work has enough time and does the job well, the DNA "hits" will disappear, and no tumors will begin to form.

Yet your body has learned through long experience that immediate repair of damaged cellular tissue only begins the process of restoration. Melanocytes are activated, spreading pigment all across the skin to protect against further UV injury—what we call a tan. If the UV damage is large enough, red splotches of inflammation pop up. White cells spread from arterioles and capillaries to cart off dead cells and clear other debris, sending back signals to further increase blood flow.

That small slip on the curb damaged cells in the anterior cruciate ligament of your knee. Inflammatory cells race over, signaling nearby connective tissue cells to start gearing up to divide. They will make new cells to replace those lost. Your walk invigorated leg muscles, many of which start to grow new actin and myosin proteins, which will make your muscles stronger and quicker.

All this repair and restoration occurs passively and unconsciously as you sit and talk with your colleague. Yet social rest and rejuvenation are also taking place. She tells you what's going on in her life. You tell her about yours. As you chat, you realize how many different connections you have with her. Your emotional responses to your children are similar; you both have young daughters who drive you nuts. And some of the ways she's handled her kid might work with yours. As you talk about your children's behavior you smell the different dishes flying by on the waiters' outstretched arms, and blood flow flies to your gut. These absorptive cells will die within the day, putting out signals that they too need replacement.

When you finally get to sleep that night, your brain will process all the information of the meal, the thousand smells from walking past restaurants and coffeehouses, the pale pink shirt of your waiter, the grinding noise of an accelerating motorbike that reminded you of your dental appointment next week. And you'll remember the story your colleague told you about her young daughter's birthday party and what her daughter's face looked like when she was told whom she could invite.

During your sleep, according to the research of Giulio Tononi and Chiara Cirelli of the University of Wisconsin, your brain will then "dump" much of this information. What is left will be the gist, the processed experiences encoded in your brain. When you wake you'll recognize that your newly married niece was thinking of moving into your friend's neighborhood and will be very interested to know a new elementary school is being built right there.

Sleep is useful in other ways. When we're young, about a fifth of sleep is spent in deep sleep, a stage of sleep that is almost a state of consciousness yet is so close to coma you will not remember any of it. In deep sleep your body produces growth hormone. Growth hormone literally makes things grow, redirecting hormones and other information molecules to subtly rebuild and reshape your body. Your muscles and connective tissue will reform, helping make you stronger and trimmer. Your skin may become less slack.

Unfortunately, deep sleep declines with age, and so does growth hormone production. Deep sleep declines more in men than women, as healthy men by their midsixties may not experience any deep sleep at all (No, life is not fair) To compensate, some people take growth hormone injections, which can cost tens of thousands of dollars a year. But if you know simple techniques described in this book, you can increase your amount of deep sleep without any drugs.

In this book you will learn several different, easy ways to sleep well. Included will be your very own sleep makeover. Sleep is, after all, a critical form of passive rest. But great advantages also come with active rest, of which social rest is just one useful type.

Active Rest

Lots of people tell me that they just don't have time to rest. Others tell me that rest is a stupid kind of laziness. Many people think rest is boring. Others think rest is something that they are forced to do when they just can't keep going, as when running a marathon—either on their feet or on the job. To huge numbers the image of rest is sitting incapacitated in a chair, staring into space, physically or mentally exhausted. Rest, for many of them, means doing nothing.

Yet if you know anything about physics, you'll quickly recognize that "nothing" is never nothing, or true emptiness. From "nothing" came the Big Bang, which created the universe and our world. The "nothing" of empty space is filled with dark energy and dark matter, which under present scenarios constitute 96 percent of the mass of our universe.

That's right. Dark energy and dark matter constitute most of the stuff of our universe. We don't know much about either of them, but we do know dark energy and dark matter are crucial to our existence.

For a lot of us, rest is a lot like dark energy and dark matter. It's critically important, but we don't pay much attention to it. Yet like dark energy, rest is fundamentally active. And unlike dark energy and dark matter, we can use rest whenever we want.

With active rest, you get to rewire and rebuild your body. But more powerfully, this book gives you an easy, thirty-day plan that provides you at least one simple rest technique *every day.* In these thirty days you will learn to consciously and thoughtfully rest in ways that will improve your health, your mood, your social connectedness, your vitality, and your creativity. Most important, you'll have more fun.

Part of the fun lies in actively directing what happens to your body and brain. And it doesn't have to take much time. Most of the active rest techniques taught in this book can be accomplished with practice in a minute or less. It's fun to feel more active and alert after thirty or sixty seconds, but even more useful, people tell me, are the techniques in this book that teach you to remain calm and relaxed in the middle of a maelstrom.

Part of the fun of learning active rest techniques is realizing how much more effectively you can focus and concentrate. Many of life's peak experiences occur when you're totally engrossed in what you're doing. Putting together the different active rest techniques allows you both to pay greater attention and to relax, to feel better, and to get *more done with less effort,* as you program your body and mind to tune themselves to greater pleasure and overall effectiveness. Rest can be sexy, fulfilling, calming, and exciting.

Perhaps rest is undervalued because we have not had a typology, a way of naming the different kinds of active rest. Here are four different kinds; hopefully future research will provide more.

The Four Kinds of Active Rest

Mental rest means focusing intelligently on your environment in a way that's rejuvenating. Techniques of mental rest give you the ability to obtain calm and relaxed concentration quickly and effectively and to become relaxed and focused anytime and anywhere. Mental rest allows for greater concentration, awareness, and achievement.

Social rest means using the power of social connectedness to relax and rejuvenate. Social rest provides a feeling of belonging and togetherness that prevents heart disease and cancer while providing you pleasure, purpose, and love. Your walk to lunch with your colleague is a small example of social rest, which quite literally can save your life. Most research studies argue social connectedness is at least as important to health as controlling hypertension and not smoking. Recent studies of the world's longest-lived populations show that social rest is probably far more important to overall survival than most researchers think. (It even works well in the frequently workaholic United States, where some socially well connected subpopulations have a life expectancy of well over ninety years.)

Spiritual rest is the practice of connecting with things larger and greater than ourselves, which provides fellowship and meaning in life—factors people hunger for like food. Spiritual rest can create a sense of internal balance and personal security while providing comfort where none appears to exist.

Physical rest, by focusing your body and its simplest physiological processes, provokes calm, relaxation, mental alertness, and surprisingly better health. It can be practiced almost anywhere, anytime— like most active rest techniques.

Even though sleep is a passive form of rest, it is critically important. Sleep right, and you will control your weight, learn better, remember more, look better, and feel awake and alert in ways you may never have experienced before. That's why this book starts with your very own sleep makeover.

Yet to form an idea of just how enjoyable active rest can be, consider sex.

Sex as Social Rest

Americans talk, dream, think, fantasize, and joke about sex a lot more than they engage in it. That's sad. Sex can produce many terrific results, as sex is a particularly pleasurable form of social rest.

As an expert on body clocks, I can assure you that sex is generally better at certain times of the day, and it is far more enjoyable when the partners are rested. Do you want to make love *only* at the end of the day, when you're plain exhausted?

Hopefully not, for sex can be powerfully restful. Even for those who prize their gymnastic prowess, sexual relations create great physical relaxation. The French call the male response to sexual fulfillment *le petit mort*, which many of their partners think is exactly right—the little death. Watching your spouse go from full arousal to snoring is not what most women (or men) crave.

Sex is also mentally restful. Whether fully aroused or absolutely sated, people describe their minds during and after sex as happy, free, and rejuvenated. Yet when used as a conscious technique of social rest, sex can provide a powerful social connection.

Sex as a form of rest also points out some of the many advantages of active rest. Active rest techniques are simple. They usually involve things you inherently know how to do. Active rest techniques are easy to combine so that social rest, physical rest, and mental rest can be achieved through the same uncomplicated actions. They also can have far-reaching effects on your health, longevity, and pleasure though they may sometimes seem the simplest things in the world.

Done correctly, rest techniques can even feel rhythmic and musical. You can learn to put them together in ways that create your own daily music, with its own tempo, meter, and rhythm. When you practice active rest, other parts of your life can also become more musical. And you yourself will decide when and how to use the many different rest techniques. Making your own daily music, you can also have others enjoyably play their personal music along with yours. It's all part of the Power of Rest.

The Power of Rest Plan

This book is separated into two parts. The first part is a thirty-day Power of Rest Plan in which you learn new rest techniques and skills *every day*.

Part I: The 30-Day Plan to Reset Body and Mind

Days 1-7: Your Sleep Makeover

In the first week of this thirty-day plan, you first look at how you actually sleep and then learn really simple ways to fall asleep—and stay asleep. During sleep, your brain takes what you learned the previous waking day and processes it, preparing you to begin the next day with a newly rewired nervous system. If you could remember what you did during the different phases of sleep, you might think you had lived several different lifetimes.

Sleeping should be easy and fun, though often it is neither. So each day of this first week you learn a quick, easy technique to relax, calm, and then prepare yourself for powerful, rejuvenating, creative sleep. After all, you need to sleep well so that you can fully enjoy the techniques of active rest.

Days 8-11: Physical Rest

In these four days you will learn six separate easy and quick techniques that will let you physically rest anytime, anywhere—and do it fast. You will learn to focus on your body in entirely new ways that improve physical and mental health. With some practice, you'll also figure out how to calm yourself in the middle of chaos, with the added advantage that no one may have noticed just how you accomplished it.

Days 12-15: Mental Rest

Over these next four days, you will pick up five different skills that will help you achieve mental rest. You should then be able to mentally rest in the middle of a boisterous business meeting, waiting on an endless airport security line, or trapped in your car in paralyzed traffic. With these different techniques to relax and rejuvenate your mind, you will begin a process that can make you a more centered, balanced, and potentially more interesting person than you thought possible.

Days 16–20: Social Rest

Over the next five days you will learn six quick techniques for social rest, ways you can connect with people you love *and* people you want to know better, all the time providing yourself with greater feelings of security and stability. People want to feel safe, and social rest can make them feel safe. Social rest can also improve your ability to stay healthy, prevent heart disease, stroke, and cancer, giving you a much better shot at a long, healthy life. It's also a large part of what makes life fun.

Days 21–23: Spiritual Rest

Here you will pick up five different techniques to obtain spiritual rest. Done quickly or at leisure, they will let you see the world in different ways while providing perspectives on life you may have not yet imagined. There are few gifts as interesting as seeing the world anew.

Days 24–27: Rest at Home

Rest and rest opportunities change as we age. Living conditions also change. Here you will learn four techniques you can use to rest as a family, or alone, that combine the advantages of physical, social, and mental rest. Rest, like life itself, should be musical. Resting at home can provide its own special kind of music, sometimes loud, sometimes so quiet you recognize the melody only long afterward. Seeing clearly just what you've accomplished is only one of the pleasures of learning to rest at home.

Days 28-30: Rest at Work

Most people haven't found any (1) time to rest at work, (2) places to rest at work, or (3) ways to rest at work. Here you'll learn four separate techniques you can do even when your boss is looking straight at you. Rest at work is critically important, as it not only provides spiritual sustenance but can also give you solace when you are pretty sure there's none to be had.

In just thirty days you will have learned more than thirty different techniques to rest, all of them ways to use your body and mind to renew and restore you. Most of these techniques can be accomplished in less than a minute. Most of them are truly simple and easy. Most can be used anytime, anywhere, to rest and restore. Some can be done to give you quick shots of power and support, what we call Power-Ups. Yet much of the pleasure comes from putting them together—sometimes through your very own creations.

Part II: Putting It Together—Making Life Musical

Now that you have all these new skills for active rest, it's important to place them correctly and make them work well in all the aspects of your daily life. With proper rest, you can make your life musical. In the first chapter, "Sequencing," you'll learn how to deal with multitasking and boredom using the concept of flow.

Lots of people know multitasking is a bad idea, but they enjoy it too much to change. Many teenagers love the four-ring circus of posting on Facebook, playing video games, texting friends on their phones, and watching TV. You'll learn what to do when multitasking is unavoidable. Yet while learning how to sequence multitasking is important, it is not nearly as important as understanding flow.

Never heard of flow? You're not alone. Mihaly Csikszentmihalyi's concept for obtaining peak experiences has been around for decades and has been cribbed by virtually every guru of positive thinking and psychology, usually without acknowledgment. Flow experiences involve these different elements:

You're engrossed in what you're doing.

Your sense of time changes; often you don't notice time passing at all.

There's a challenge you're working on.

You're using different skills to meet that challenge.

You're getting feedback as to how well your skills are working.

A standard, simple flow experience is a game like tennis. If you're enjoying the game, you're engrossed, time flies, you're trying to win the game, and you can clearly see if your skills are improving. As you become familiar with the active rest techniques, you will take your favorites and put them to use turning some of your most ordinary activities into flow experiences. And with a little practice, the rest techniques themselves will become flow experiences. That means that feared periods of boredom, listlessness, and frustration can be changed into times of focus, concentration, self-awareness, and achievement. It really does pay to know how to rest well.

The next chapter, "'I Have to Do It': Eliminating the Required, Doing the Essential," looks at how you really spend your day. Are you doing too much? Too little? Prioritizing and finding perspective in life are not easy, just necessary. Here you'll look at your weekdays and weekends, checking out what you do and what you really care about. What you want is to become effective rather than efficient and have more time to do what you really like—especially rest.

You want to make life rhythmic and musical. It helps that we are inherently musical. The final chapter of the book, "Tuning Your Life," shows you how to create a rhythm in your life that is pleasurable as well as productive.

The brain and body are made up of many systems. How these many systems integrate and communicate makes life possible and entertaining as well. Human life, indeed all life, operates in the nexus between energy and information.

A lot of your body's information is communicated through rhythmic activities. Many scientists, like codiscoverer of DNA's structure Francis Crick, think consciousness comes about when some of our many different brain systems start to fire synchronously. Thought really is action, the action of different brain cells talking to one another at the same time. We may like rhythm and music so much because rhythm is what our brains are made of.

Yet lots of people have a hard time making their lives rhythmic and musical. They really do not have a clue how to do it. When do I rest? When should I be engaged in activities to create flow opportunities, experiences providing for maximum performance and pleasure? How much food and what kind do I need? If rest is so important to renewing my body and my brain, how do I balance it with activities I must do to make a living and survive?

Going FAR is one way to start making life musical. It's simple, easy, powerful, and conceptually compelling. Again, FAR stands for food, activity, and rest. Food provides basic fuels and living materials. Activity is what we do. Rest is how the body, passively and actively, rebuilds itself and nourishes the mind. Going FAR is one way to start to make your life musical and sequenced. It has the innate advantage of fitting the body's powerful timing mechanisms, making your normal activities better fit your own human design.

When done right, FAR helps people feel energetic, alive, amused, and more mentally and psychologically balanced, with the added benefit of making people physically healthy while letting them control their weight. Not bad for three letters put into a simple sequence.

Of course, there are many other ways to make life musical. Some of the music comes from waking in the morning and noticing your many different powers of perception, things you will learn about when you do physical and mental rest techniques. Other music comes from time spent talking, laughing, working, and living with others, through the many different kinds of social rest. Other musical strains come from the powerful spiritual stirrings that affect people through the many different phases of their lives, means that can be developed by the simple techniques of spiritual rest. Harnessing these different kinds of music can increase pleasure and productivity as well as give you access to the secrets of the longest-lived people on Earth.

Yet going FAR provides only a simple, if powerful, rhythm to daily life. Music is much more than rhythm. Music involves key and harmony, pitch, timbre, and melody. To achieve real music, you first need to know how to rest.

Rest is powerful. Effective rest can help you succeed by making you alert, whole, productive, and happy. It can also provide meaning, making you and your life both more interesting and more complex. With this book, you can learn the rudiments of rest, techniques that you can take with you anywhere. With a little practice, you can then apply these techniques to work, love, leisure, and spirit. It's time to learn how to rest, to build inner and external strength, to feel healthy and alive, to obtain greater control of your life and consciousness, to get more time to do what you really want to do—and to do more with less.

Since so many people don't really know how to enjoy it, let's start with the great, no-longer-so-mysterious mystery of sleep.

chapter 2

SLEEP RENEWS YOU

Your One-Week Sleep Makeover

Life is rhythmic. Everything we do follows cycles of activity and rest, of which sleep is a critical part. Life involves continuous regrowth and renewal of our cells, tissues, and information systems. Sleep is a necessary part of that regrowth. Living each day according to the rules by which our bodies are designed, we gain better health and experience lives with greater psychological meaning and balance. We all need good sleep, just like we need food.

But sleep does more than provide for better health and psychological balance. For millennia people have tried to understand what sleep is for. Here are a few of the answers that have appeared only in the last few years:

1. Sleep is necessary for weight control.

2. Sleep is required for memory and learning.

3. Sleep is needed to prevent major clinical depression.

4. Sleep is required to grow new brain cells.

5. Sleep is needed to avoid colds and fight off infection.

6. Enough sleep prevents plaque from forming in your arteries, preventing heart attack and stroke.

7. Proper sleep is required to maintain and strengthen the inner clocks that regulate our lives.

Sleep, minute by minute, uses more time for most of us than any other form of rest. For our health and well-being sleep should take up about a third of our adult lives. Perhaps because our actions in sleep appear so passive and because our consciousness is turned off while we sleep, we fail to appreciate its enormous benefits. Yet if you know how to sleep right, you can learn more and remember more. You will gain a powerful aid to control your weight. You can move a long way toward the prevention of illness and halting the ravages of diabetes, heart disease, stroke, and cancer. You'll be better able to think, and think creatively, and you will begin the process of putting flow in your life, balancing activity with rest in ways you may never have imagined.

What your body and brain do during sleep is to renew themselves. If you know how to sleep, where to sleep, and when to sleep, then sleep can be more than simple and easy. It can become pleasant and creative, perhaps the most important single type of rest to rebuild and restore you.

Rest is *the* original transformative technology. It is a capacity you always have within you. If you learn to sleep right, sleep can be more than fun. Through dreams, and the full sense of alertness and awareness, sleep can become an adventure.

What Is Sleep For?

Let's look at some of the immensely useful results of getting good sleep.

1. Sleep Is Necessary for Weight Control

From 2003 to 2009 multiple population studies around the world have shown that people who sleep fewer than seven hours, especially those getting fewer than six hours of sleep, gain appreciable weight. This has naturally led to studies in which people are asked to sleep longer. In the small-scale, short-term studies done in 2008, sleeping an extra thirty to sixty minutes a night decreased weight. Some people rapidly shed ten to fifteen pounds.

An interesting question is why. Much of the research has been done with *partial* sleep deprivation. There, instead of sleeping a full seven to eight or more hours a night, research subjects are allowed only four to six hours of total sleep time. Work done by Eve van Cauter at the University of Chicago and by her colleagues in Belgium shows this partial sleep deprivation rapidly deranges glucose metabolism. Since glucose is the body's main source of fuel, and outside of starvation the *only* fuel of the brain and red blood cells, you really don't want to foul up glucose usage. Other important hormones such as ghrelin and leptin, which operate a kind of internal yin-yang determining how hungry we are, also dramatically shift with partial sleep deprivation.

Most of the studies on partial sleep deprivation have been performed on undergraduates. After a couple of weeks of their normal sleep they go to the laboratory, where they're allowed only four or six hours of sleep a night.

Within a few days they look prediabetic, improperly producing insulin even when glucose levels increase. This process, known as insulin resistance, quickly gets worse as these studies continue through separate nights. Insulin resistance has many unpleasant results, including unsightly fat deposits in the abdomen. Insulin resistance also sets people up for future diabetes and cardiovascular disease.

The good news—sleep enough, and glucose metabolism is even better controlled. You prevent a major cause of insulin resistance, a scourge of American life. You don't mess up ghrelin and leptin levels, which help tell your body how much you should eat (including how much sugar). Sleep right, and you need not get the munchies, in particular those sugar cravings so often experienced by night-shift workers and others who can never get enough sleep.

2. Sleep Is Required for Memory and Learning

The better you sleep, the more you learn. In the last ten years sleep researchers have more clearly identified which stages of sleep lead to improved learning.

Just as life is rhythmic, so is sleep. During the first part of the night, in approximately ninety-minute intervals, people cycle from light sleep to deep sleep to REM, or rapid-eye-movement, sleep. Deep sleep looks a bit like coma on traditional brain electrical studies, but during deep sleep enormous amounts of information are digested and put into usable form. Similar brain work is done during REM sleep, the only part of sleep we tend to remember at all, mainly because it includes complex dreams. If we did remember what we did and thought during our different phases of sleep, we'd probably consider each of them separate states of consciousness.

Recent animal studies provide more clues to sleep and learning. In rats, the brain enzymes necessary to lay down memory traces don't even begin to work until the animal goes to sleep.

In humans, deep sleep decreases with age, more so in men than women. As we age, REM also drops out a little bit. Your sleep make-over includes techniques that help reverse these effects of aging.

Research at Harvard has shown that short periods of daytime sleep—even as short as a six-minute nap—can improve memory. Complex decision making in particular is aided by getting a full night's sleep. Many people will tell you they often made the most important decisions in their life, like what job to pursue and whom to marry, following a good night's sleep. Many creative ideas, including great scientific breakthroughs, are created while we sleep.

German researchers did an interesting and peculiar experiment. They gave people a choice of three separate car deals. All were difficult to evaluate, but one was better—if you could figure out all the details.

They tested people during the daytime, then after a night's sleep. The participants came up with much better answers once they'd had a full night of sleep.

The Stages of Sleep

Stage 1: Light sleep. Though useful, stage 1 sleep is so light that many people fall into it and think they are awake, leading to endless couple arguments and increased accidents. Light sleep makes up 5–15 percent of normal adult sleep.

Stage 2: Marked by funky sleep spindles on an EEG, stage 2 sleep is the majority of human sleep (55–60 percent). Lots of regrowth going on.

Stages 3 and 4: Put together, they are called deep sleep. Growth hormone is produced, and lots of learning and memory building occur here. In adolescence, deep sleep is 15–20 percent of sleep, but then declines throughout adult life.

REM sleep: The most "active" part of sleep (its old name), REM involves loss of temperature control, highly complex dreams, and lots of complex learning, often done in concert with deep sleep. Makes up 22–24 percent of sleep.

If you have a tough decision to make, sleep on it.

Unfortunately, that's happening less for American adolescents. Adolescents need far more deep sleep than adults, which may be required for their brain development, as perhaps a third of their neural connections (synapses) die and become replaced during the teen years. Adolescents need nine or more hours of sleep in order to really remember and learn. These days, many American teenagers get fewer than six hours.

There's too much to do at night—instant messaging, surfing the Net, texting, games.

Another reason kids don't sleep nearly enough lies in the early start times at many schools. If your daughter is getting dressed, putting on makeup, and slurping down breakfast before waiting outdoors for her

6:45 a.m. bus, don't be surprised if she sleeps through the first hours of class. Through programs spearheaded by my teacher, Mary Carskadon of Brown University, school districts are strongly encouraged to start classes at biologically sensible times. Studies show that the students who get to school at later times are already doing better academically. More sleep means better performance. The same is true in college, as studies at the Air Force Academy demonstrated in 2008. As sleep is also necessary to control weight, increased sleep should help adolescents improve their ability to prevent obesity and type 2 diabetes, already major public health problems throughout the world.

3. Sleep Is Needed to Prevent Major Clinical Depression

When people sleep enough, their mood quickly improves. In America, the less sleep people get, the more tired and cranky they become. It's true whether the insufficient sleep occurs over days or, worse, over months or years. People feel a much greater sense of well-being when they sleep well. And getting enough sleep should also go a long way toward preventing major clinical depression, an illness whose rates have been doubling and tripling among Americans over the last thirty years.

The data showing the many mood benefits of sleep come mainly from studies of insomniacs. It's difficult to research long-term insomnia in the United States because Americans can move around a lot. In former days, a fifth of Americans moved each year.

That's not true in Switzerland. The Swiss move around very little, and each local Swiss *Amt,* or municipality, collects an extraordinary amount of information on nearly every citizen. This allows Swiss scientists to do some wonderful epidemiologic work.

Jules Angst, when he was professor of psychiatry at the University of Zurich, followed insomniacs for many years. He found that the longer they were insomniac, the more depressed they became. Angst (his name means "anxiety" in German) discovered that when people slept poorly for more than ten years, about a third then developed full-blown clinical depression.

Depression itself causes many sleep disturbances ranging from insomnia to extreme daytime sleepiness. Your one-week sleep makeover will provide you many tools to make sure your sleep is not merely sufficient but also effective.

4. Sleep Is Required to Grow New Brain Cells

When animals sleep they grow new brain cells. Most important, many of the new brain cells that divide and grow live in the hippocampus. The hippocampus is a small, deep part of the brain critical to laying down memory and learning and to expressing our emotions. As you may know from your own life experiences, the strongest memories usually are emotional ones.

Studies of new brain cell growth have been done since 2007. The work, performed in rats, has occurred at several universities, including Johns Hopkins. Though done in animals because of the deleterious effects of even small brain biopsies in people, the studies show some startling results. For over a century scientists and doctors were taught that humans and other complex animals did not grow new brain cells. The dogma was propagated by Pierre Paul Broca, one of the most important brain scientists of the nineteenth century and one of the first to exhaustively map the human brain.

Broca looked hard to find any new brain cells. Every time he looked, he could not find any. Every other type of human cell seemed to die and reproduce. Not brain cells. Broca decreed brain cells were different, and that was that.

Considering that your brain is 90 percent of its adult size by the time you're two years old, that means that the number of brain cells you've got soon after birth is pretty much all you will ever possess. (The brain is strange in a lot of other ways. Babies' eyes look really large compared to their heads because our eyes don't grow as we mature. The optic nerve is part of the brain, after all.) Yet recent studies in rats show they do make new brain cells every night, primarily in the hippocampus. If the rats are not allowed to sleep, they don't create new cells. Broca was wrong—we create new brain cells, and we do it in sleep.

5. Sleep Is Needed to Avoid Colds and Fight Off Infection

Plenty of research shows that animals fight off bacteria and viruses far more effectively when they get normal sleep. In animals, insufficient sleep often provokes much higher rates of fatality from bacterial infections. Data from Carnegie Mellon University published in 2009 shows that sleeping well also defeats one of our most highly evolved, ubiquitous enemies—the common cold.

Adults filled out sleep records for two weeks then voluntarily allowed technicians to directly instill cold viruses into their nostrils (one reason why people get paid to become research subjects). The baseline group for the study was made up of those sleeping eight hours or more a night. Those who slept fewer than seven hours got three times the number of colds.

How efficiently they slept also powerfully affected their ability to fight off a cold. Those with a self-reported sleep efficiency of 92 percent or less had nearly six times the number of colds of those who slept 98 percent or more of the night.

Sleep efficiency? Fortunately, it is a real number that can be studied. Sleep efficiency is officially defined as

Time asleep

Divided by

Time in bed while trying to sleep

If you look at surveys, huge portions of U.S. and European populations report sleep efficiencies of less than 92 percent. For an eight-hour night, that means being awake a total of thirty-eight minutes or more—not much. And these are subjective numbers. Except for insomniacs, subjective sleep numbers are often higher than what is reported when people sleep in a laboratory, where electroencephalographic monitors can determine how many seconds your brain actually sleeps.

A subjective sleep efficiency of 92 percent is actually rather good. That's something to think about when you're trying to make it through work during a long winter afternoon, coughing, sniffling, aching, and tired, wondering how you will ever finish the workday and do all the stuff come evening that your kids and spouse expect from you. Rest is more than restoration. Rest is on the front lines of your infection defense system. It keeps your immune system going. Rest prevents infection. Rest well, stay well.

6. Enough Sleep Prevents Plaque from Forming in Your Arteries, Preventing Heart Attack and Stroke

You want to prevent plaque buildup wherever it takes place. Bacterial plaque damages gums, but arterial plaque can kill you. Arterial plaque is the basis of the majority of heart attacks and strokes, and fortunately for us, good sleep can prevent its even getting started.

Plaque is the goop that clogs your arteries. Plaque begins to form when inflammatory white cells move into arterial walls and stuff themselves with fat. Soon the white cells morph into weird creatures called foam cells. Lots of foam cells become lots of plaque, until your arteries narrow and narrow, the opening between the walls no longer circular but thinned and lopsided.

Yet the majority of heart attacks occur, not in people whose arterial walls are plagued with plaque, but in those whose arterial openings are narrowed just a little or hardly at all. On standard cardiac catheterization, these people look okay. But bad events can happen. That wall-infesting plaque can burst out into the artery, causing clots to form and arteries to spasm. If the plaque pops out in the wrong place, like in some tiny coronary arteriole supplying the pacemaker cells of your heart, you may suddenly die, even if your arteries looked nice and normal on a coronary catheterization. Unfortunately, such events occur frequently. Cardiologists may stent arteries that are 70 percent or more occluded, but the real damage to the public health occurs with less-narrowed arteries.

So if you really want to prevent plaque formation, you want to sleep well. A study from the University of Chicago published in 2009 followed healthy men and women in their midthirties to early fifties over a period of five years. Somewhat to the authors' surprise it turned out that if you sleep less you make *lots* more plaque, enough to produce major coronary narrowing. People who slept on average five or fewer hours each night experienced dramatically increased coronary narrowing, but plaque formation was notably faster even in those sleeping fewer than seven hours a night. Most American working women say they are now sleeping around six and a half hours a night.

And this study looked not at self-reported sleep, but at something close to actual sleep time as recorded in the sleep lab. The study participants wore actigraphs, watchlike gizmos that check and record how much you move during the day and night. Since people generally move a lot less during sleep than when awake, actigraphs can provide a fairly good approximation of total sleep time. The young to middle-aged men and women in the Chicago study were sleeping a good deal *less* than what they thought they slept. Certainly most people don't voluntarily sleep five hours a night, no matter what problems they have with kids and multiple jobs.

Want clean, plaque-free arteries? Get better sleep.

7. Proper Sleep Is Required to Maintain and Strengthen the Inner Clocks That Regulate Our Lives

You can get healthy sleep at different periods of the night—and day. You just have to know when.

Huge numbers of people, including many sleep docs, think you need to get all your sleep in one big, uninterrupted clump at night. This concern for getting all-in-one sleep has itself become a major cause of insomnia, particularly as people whittle down the time they allow for sleep and rest. Concern about getting perfect sleep produces the common problem clinicians call psychophysiologic insomnia, which is failure to sleep because you're worried you're not getting enough sleep. Psychophysiologic insomnia is common in working mothers and professionals.

The truth is, sleep can be gotten at several natural periods of day and night. Yes, sleep efficiency, getting lots of sleep during the periods you want to sleep, certainly matters. But natural sleep is not required in one phase at night or even in two nighttime phases. Sleep can be naturally performed two or three times during the twenty-four-hour day. Normal human sleep is probably triphasic.

The data on natural sleep goes way, way back, into preindustrial times. In days when candles were the main form of nighttime illumination, many people slept nine and a half hours or more each night. Studies of historical diaries show that many folks also regularly woke up in the middle of the night. Then they'd talk to the others in the room or in their bed (beds were expensive, and families often slept together), get up from bed and do housework, or simply muse about their lives before returning to sleep.

Life even before candles was different in several other ways. Recent work at the National Institutes of Mental Health attempted a laboratory version of how our cave-dwelling ancestors slept. Research participants went into a specially created environment without sunlight. There they were asked to sleep through each "night," from artificial sundown to artificial sunrise.

Thomas Wehr reported some of the results at the Associated Professional Sleep Societies' annual meeting, which spreads a sleepy pall on another American city every spring. Many of the study participants, he said, got up in the middle of the night to ponder their dreams. Quite a few described mystical experiences during the night.

When the study ended, some participants wanted it to continue. They had never felt so alert and awake. Others described a sense of personal wholeness and fullness uncannily similar to how members of hunter-gatherer tribes describe their personal relationship with the natural world.

Lots of people wake in the middle of the night, think about their dreams, and go back to normal sleep. Many, many others routinely go to sleep during the daytime. Naps, or siestas, are commonly performed by over a billion humans across the planet. In the past, before industrialization, they were routine for nearly everybody. Why? We're programmed that way.

Our inner body clocks are a major factor controlling when we sleep. The Romans said "time rules life." Indeed, inner biological time is the armature of life.

During the afternoon, most of us get sleepy. That's normal. It follows your inner body core temperature. When your core body temperature is going up, you're more alert. When it's going down, you get sleepy.

In the afternoon, body temperature gets pretty flat. That sets us up, at least lots of us, to sleep.

Naps have many uses. They can get you through a long workday, refresh you on the weekend when you try to overcome normal sleep deprivation, or just give you a different perspective on a persnickety problem. Even brief naps can improve memory. (There's more about naps and near naps in chapter 4, on mental rest.)

It's certainly fine if you want to get all your nighttime sleep in one fell swoop. But it's not required and certainly not essential. It may be perfectly okay for you to get up at night, go to the bathroom, come back and read a few minutes before sleep (without looking at the clock), or take a brief nap during the afternoon. It all depends on your own personal biological clock. What does matter is the regularity of your pattern, when you wake and when you sleep.

Your clocks are regular in their timing for a reason. Life is rhythmic. So is sleep. To know when to sleep, it pays to know a little about your own inner music.

Rhythm, Sleep, and Music—Your Body Clocks

Did you ever wonder why people in virtually all human societies enjoy music? From a biological survival standpoint, music does look pretty superfluous. Despite their many obvious talents, do Tom Jones, Barry Manilow, and Guns N' Roses really promote human evolution?

Evolutionary biologists would probably say, well, yes. Look at birds. Birds sing, often with profoundly romantic intent. Male birds that are musically gifted demonstrate to females not just originality and cognitive ability but also firm evidence that their bodies are in good health. They also proclaim through their highly individual songs that they will make good mates. The more musical songbird guys have a better chance of getting the most desirable gals.

For birds, it really does work that way. Human evidence, such as the memoirs of rock, pop, and jazz stars, might indicate that such evolutionary theories also describe human behavior; the popular singers do get the girls—or guys. Yet there is a simpler explanation.

Life is timed. Your life is rhythmic.

In human life virtually all biological processes run in cycles. Some of these cycles are quick, like the millisecond flows of neurotransmitters between nerve cells. Other human cycles take many years. One example is human maturation, moving from birth through puberty to middle age and beyond. Quite a few cycles occur around periods of a month, as in the lunar cycle of menstruation.

One reason music may be so natural to us is because our cells and organs communicate with each other rhythmically. We speak rhythmically, a rising and falling burst of syllables distinct to every language and speaker. And our days flow through the rhythm of our actions, from the times we wake up to the times we eat, work, move, and sleep.

Just as life is rhythmic, so is rest. Rest well, using sleep—as well as the various forms of active rest—as a modulated, rhythmic part of daily life, and your own days can become musical.

Amid the many human biological rhythms, the most studied are those of day and night. Raptors that soar through the skies as well as creatures that live five thousand meters (three miles) deep inside the Pacific Trench all operate according to twenty-four-hour rhythms, called *circadian rhythms*—Latin for "about a day."

Circadian rhythms are a basic design element of life on Earth. All living beings on this planet have many internal ways of adjusting to the rhythm of day and night—and the different environments such changes produce. Humans have internalized those times into our basic makeup.

Scientists at Colorado State University studying human genes reported in 2008 that about 98 percent or more of our genes operate according to twenty-four-hour patterns. Life is timed, and sleep, like other forms of rest, also follows circadian rhythms. Circadian rhythms powerfully affect our behavior and our performance. Want to set a personal record in an athletic event? Try the sport during the late afternoon or early evening. Want to lose weight? Consider a big breakfast. Want to test well after cramming, using your short-term memory? Take that exam in the morning. Body clocks often influence when we are best at doing things, whether it's thinking, eating, losing weight, sleeping, or, for maximum memory retention, reading this book (try the evening; long-term memory is better then).

Except human nights are not created equal. We may all live by twenty-four-hour clocks, but every one of us has our own personal version, with different start and finish times.

Larks and Owls

I'm a morning person. My body generally wakes up around or just before my clock radio goes off, a performance achieved daily by about a third of Americans. In my case wake-up is 6:00 a.m., an hour many of my friends consider the middle of the night.

Those friends sometimes tell me they've won a victory when they keep me up past 11:00 or midnight. They've watched my eyes get glassy as the clock moves toward 10:00 p.m., my habitual sleep time. I can easily keep myself up drinking tea or eating lots of chocolate but my body will still want to get up at 6:00 a.m. If I go to bed too late, my next day will not be pleasant, filled with that cranky sluggishness and brain slowness many of my patients complain about before I treat them.

Owls and larks, night and morning people, are born to their rhythmic ways. There are plenty of genes that determine which section of the twenty-four-hour divide you fall on.

Since our work world caters primarily to larks, being an owl creates lots of complications. Many of my owl colleagues tell me that they tend to make friends among other owls, and that they did so especially during their teens and twenties (remember, teenagers have bodies that go to bed later and sleep longer—at least that's what they should do). Not infrequently, owls mate with other owls they meet at late-night parties after everyone else is heading home or has already arrived there.

Friendship, love, and jobs all may become owl-centric. For years and years I searched for a lark who was a jazz musician. I finally found one—a fellow whose preferred wake-up time was 7:00 a.m.

He loved jazz. He loved making music, and he loved his musician friends. But he quit professional jazz after two years.

Most of the population is neither lark nor owl. They're in between, called sparrows by some, hummingbirds by others. They are the people who are catered to by television news, theater, and NFL football, falling asleep between 10:30 and 11:30 p.m. and rising before 8:00 a.m.

To find out whether you are lark, owl, or sparrow, go to the full Biological Time Test in chapter 7. Have your girlfriend or boyfriend look at it too. Owl and lark marriages do have somewhat higher divorce rates.

But if you want to know when to go to sleep and get up, try this quick part of the Biological Time Test.

You're on the vacation of a lifetime. It's for as long as you want, with no worries and no responsibilities. You've got more money than you'll ever need, and you can do precisely what you like every minute of the day.
What time would you go to bed?

Between 8:00 and 9:00 p.m.?
Between 9:00 and 10:00 p.m.?
Between 10:00 and 11:00 p.m.?
Between 11:00 p.m. and midnight?
Between midnight and 1:00 a.m.?
Between 1:00 and 2:00 a.m.?
Between 2:00 and 3:00 a.m.?

Write down your preferred sleep start time here: _____

You're enjoying the best vacation of your life. You are doing everything you want to do, how and when you want to do it. Getting up feeling completely alert and alive is important to you, and you give yourself as much rest as you need.
What time would you wake up?

Between 5:00 and 6:00 a.m.?
Between 6:00 and 7:00 a.m.?
Between 7:00 and 8:00 a.m.?
Between 8:00 and 9:00 a.m.?
Between 9:00 and 10:00 a.m.?
Between 10:00 and 11:00 a.m.?
Between 11:00 a.m. and noon?
Between noon and 1:00 p.m.?
Between 1:00 and 2:00 p.m.?

Write down your preferred wake time here: _____

All right, now you have a good idea of *when* you biologically should sleep. But before we start our one-week sleep makeover, you need to ask one more question.

How Much Sleep Do I Need?

How much sleep do you need? Enough to feel rested, alert, and alive. Yet it's astounding how people regard sleep and rest as a waste. Many CEOs, doctors, or accountants point to their lack of sleep as a sign of hard work, conscientiousness, superior mental conditioning, or macho superiority. Many years ago I heard about one cardiologist who boasted triumphantly to another of the changes he wrought once he took over a different university cardiology department: "I used to sleep three hours a night. Now I sleep zero hours!"

Electric lights, telephones, and the Internet all at their time proved to be transformative technologies. Well, rest too is a transformative technology, and an extremely powerful one. What is more, rest is at your beck and call. You rebuild your proteins, renew your cells, create new brain areas, rewire, reconfigure, and renew yourself as you rest—if you do it right.

It turns out the amount of sleep people need to feel rested varies a lot. No single number fits everybody. The average forty-two-year-old female does not require 8.2 hours of sleep or 7.3 hours. Your sleep need is how much sleep time *your* body needs.

A 2008 study purported to show that survival was closely related to the number of hours slept. Those who slept more than eight hours, especially those who slept nine hours or more per night, died at faster rates than those who slept seven to eight hours each night.

One woman called me after reading the article, not realizing that the group of long sleepers included a lot of people who already had chronic medical conditions, a major reason for their increased mortality. She had told her more-than-seventy-year-old husband he was sleeping too much. She carefully monitored him, allotting him his "maximum survival" eight hours in bed. He promptly became frustrated, tired, cranky, and really, really sleepy. Her husband just needed his normal allotment of sleep. Some people's restoration time is longer than others.

Your personal sleep need, like your hair color, has a strong genetic component. Studies done in 2009 of a mother and daughter with a rare version of the DEC2 gene, which does something still unknown to circadian rhythms, showed that the two slept very normally and restfully from 10:00 p.m. to 4:00 a.m.—six hours a night. That was their normal sleep.

In my college years I was astounded by an upperclassman who reveled in sleeping at least ten hours every night.

"Don't you think that's too much?" I asked.

"No. I love sleeping. It's great. I feel wonderful when I get up."

He was a funny, witty guy, but we worried about his capacity for striving. Amherst College was not made up of people who wasted ten hours a night on sleep.

His early career—marking student papers for professors rather than moving on like the rest of us to professional or graduate school—appeared the harbinger of a less-than-meteoric future. In the end he became a professor at a famous business school, then an entrepreneur, eventually a financial whiz and philanthropist. Being a long sleeper did not handicap him.

My brilliant colleague Lee (not his real name) represents the opposite end of the sleep need spectrum. He needs two to three hours a night to feel rested, though with age he sometimes prefers four.

When he was writing a book with a colleague, they were working hard and long to meet a not-too-distant deadline. Once their writing session finished after 1:00 a.m., they both went to sleep in Lee's apartment. Lee woke his friend to start working again. It was 4:00 a.m.

"You lunatic!" his friend shouted. He pulled the covers up and went back to sleep.

Lee kept going, of course. When young, he was able to work the thirty-six-hour shifts of medical internship without a great sense of fatigue. Most nights he would get perhaps one hour of sleep, but he found that amount quite refreshing. Often he was up for rounds each morning bright-eyed and earnest while others could hardly keep their eyes open. He told me the other interns hated him.

How much sleep do *you* personally need to wake up and see the world with happy eyes, excited you're still alive and ready to learn more about rest?

To answer, consider the weekend.

How Much Sleep Is Enough: The Wonders of Weekends

Human beings did not evolve with the weekend, but for lots of Americans it would be okay if we had. After long workweeks spent running around, breathlessly trying to get everything done, we really look forward to the weekend.

One of the reasons the weekend is so popular is because many of us believe weekends are when we will get to do more of what we like to do. Unfortunately, that's not always true. One of the major reasons weekends are treasured is because weekends are when we actually get to rest. Long lunches with friends. Reading in bed. Sunday brunches. Long telephone calls to Mom and the kids. And, best of all to many of us, sleeping in.

Since the average working woman in America is getting perhaps six and a half hours of sleep each night (and wants more), the main period for her recovery sleep occurs during the weekend. Though some of that additional sleep takes place during naps, the large majority occurs by sleeping in.

Few recognize there are perils to sleeping in. Many people sleep late on Saturday and Sunday mornings, and then on Sunday evening realize, with occasional dread, that they've got to get up early the next morning for work. Many will then try to go to bed at the same hour they used during the weekdays. Unfortunately, many do not succeed.

The result—Monday morning is literally the killer time in America. In several U.S. populations, death from heart disease rises fivefold.

Aided by the biological clock dislocations brought about by weekend activities, Monday morning is the national peak time of death.

Fortunately, this excess mortality is easily prevented. Resting makes you feel better. It also makes you live longer.

Weekend Sleep

For you, weekends may be just another day for shift work or a whirl-wind process getting the kids to soccer practice, band practice, and music practice, plus yourself to the hairdresser between shopping for essentials. Now, I want you to think of those weekends when you actually had enough time to rest the way you wanted.

Try to remember those rare, relaxed weekends. If you can, recall times you went to bed on calm, pleasant Friday and Saturday nights. Next, try to consider the times the next morning you actually got up, not just to let the dog out but when you really arose, ready to start the day.

How many hours did you sleep those Friday and Saturday nights? Write that number here: _____

Next, add in the hours spent in naps Saturday and Sunday during those calm, relaxed weekend days. Write that number here: _____

Now, add up all those hours, the times you slept at night and the hours spent napping. Write that total number of hours here: _____

Next, divide it by two, and write down the result: _____ Sleep Need—The Weekend Sleep Method

That's a fair estimate of the number of hours you need to sleep— arrived at by the weekend sleep method.

Does the number seem big? Recognize that it probably includes some period for recovery sleep, attempts your body is making to get enough rest time to rebuild itself. Still, this number of hours is probably closer to your real sleep need than what you normally get on weekday nights.

If this sleep need number still looks too large, remember a time, perhaps long ago, when you had a real vacation. I don't mean a vacation in which you witnessed thirteen European countries through a tour bus window or rushed to see the relatives Christmas Eve before running back to tote up your taxes on December 31 or tried to make the kids happy by visiting every national park in Utah. Try to remember a real vacation, a vacation in which you rested, and at the end of which you felt refreshed and relaxed.

Now, consider the last few days of that vacation, when you really felt rested and had time to sleep. How much time asleep, including nights and naps, were you getting those last days?

Write that number of hours here:
_____ Sleep Need—Vacation Method

Do your hours of sleep need appear similar when estimating by both the weekend and vacation methods? If not, don't fret. Lots of people sleep far more on vacations than they do on "relaxed" weekends, while a few do the opposite.

If the numbers are not the same, do this: take the number derived from the weekend method, and add that to the number of hours of the vacation method. Once you have that total number, divide by two.

Write that number here:
_____ Sleep Need—Averaged Estimate

This is another approximation of how much time your body is telling you that you need to sleep.

Is the number still too big? Maybe way too big? There's no way you'll ever get that amount of time and perform the duties you absolutely must perform? If that's the case, you might want to take a peek right now at chapter 10, "'I Have to Do It': Eliminating the Required; Doing the Essential." If family demands and economic survival still require you to get less sleep than your estimated sleep need, decrease your sleep need by a half hour.

You've now got a fair idea of how much time your body needs to sleep. For most, it will be somewhere between seven and a half and nine and a half hours in your bed, with perhaps 85 to 94 percent of that time actually asleep. (Don't worry; ways to increase sleep efficiency are part of this sleep makeover.)

You also now know your body's preferred time for finding your bed or futon and starting to sleep. Do these times line up with the actual times you go to bed and get up?

They don't? It's time to get those hours aligned. Now's the chance to take the time you need for sleep.

Consider what might happen if you do give yourself enough time to sleep each day. You might learn better. You'll have an easier time losing weight. Your memory should improve. You can help prevent infection, heart attack, and stroke. Your mood might get much better. With a little practice, you may start to feel really good.

Think about what sleep can do for you. Rest is restoration and rebuilding. A lot of that takes place during sleep.

Okay, you're almost ready to start your sleep makeover. But we do need to consider a special case first—adolescents.

The Strange Case of Adolescence

Most people do not allocate as much time in bed as their sleep needs would dictate, but adolescence is a truly different experience. Part of our human design is that when we are teenagers we go to bed later, wake up later, and need to sleep *longer*.

If you want to know the probable cause for many of these anomalies, think brain development. According to the work of Irwin Feinberg, during puberty about 30 to 40 percent of synaptic connections, the physical bonds between nerve cells, go kaput. Then they rebuild. As puberty begins, you quickly develop a very changed, rather disorganized brain. It will take some people several years to reorganize that brain. Many of the connections the brain then makes are wholly new.

And unlike most brain changes in adults, these are changes you literally see. Adolescents behave differently. Their bodies fill up, grow longer and larger. And the process takes years. The frontal lobes of adolescents, critical parts of the brain involved in planning and judgment, may not mature until kids reach their early twenties. So there are good biological reasons adolescents get into more traffic accidents, become involved in bizarre campus rituals like hazing, and readily volunteer for hazardous military action.

Though youthful risk taking and questionable judgment occur naturally, often the end results are worse than they need to be because adolescents rarely give themselves enough time to rest. Even if they need on average nine and a half hours of sleep to do marginally well on their SATs, many adolescents would rather allocate six hours or less to sleep. They prefer nights simultaneously spent watching TV, playing video games, listening to iPods, and instant messaging while ingesting junk food and caffeinated sodas, the entire performance vastly preferred to anything resembling homework (boring but doable between instant messages) or sleep (really boring). That so many of them sleep through morning classes in both high school and university is no surprise. U.S. high school students' low international rankings on measures of learning and thinking are not surprising either.

Adults too don't like to spend much time sleeping. Why would you want to sleep when you could watch late-night television, Monday night football, weight-loss infomercials, and really great 1950s TV reruns?

You now know why you must sleep. You know your preferred times to go to bed and get up, and how many hours of sleep you really need.

So it's time. Now you can start your one-week sleep makeover: seven days to achieve a different kind of rest.

SLEEP MAKEOVER DAY 1: WAKING UP AT YOUR PREFERRED WAKE TIME

Waking up is hard to do. It's best to do it when your body really wants it.

When to Wake Up

At your preferred wake time, determined above.

How to Wake Up

Buy some form of alarm. The alarm should be very consistent and easily workable. Clock radios are generally effective. They allow you to wake to anything you enjoy listening to, whether it's Aretha Franklin, the BBC, National Public Radio, CBC, or morning prayer. Music is a nice wake-up vehicle, as it then sets up an activating, natural rhythm to your morning.

Other highly effective alarms include your husband, wife, or partner, if that person is time reliable and can be expected to wake you comfortably and affectionately; and loud, bell-like, pleasantly pitched chimes. In a pinch, buzzing and vibrating alarms can be used if you are someone who simply does not and cannot wake up.

Less Convenient Alarms

Alarms less efficient but often insistent include cats and dogs. Though all animals are circadian in their rhythms, cats and dogs have rather different twenty-four-hour rhythms than we do. Cats and dogs like to sleep a lot more than us, usually ten to twelve or more hours a day, and they prefer a lot of that sleep to occur during the daytime.

Cats in particular seem to obtain great pleasure from dive-bombing your head, jumping on your chest, or simply meowing until you pay them enough personal attention. Their preferred time for waking varies greatly but often starts in the middle of the night. Some cats will irregularly wake you at 3:00 or 4:00 or 7:00 a.m., according to rules and regulations known only to them.

Cats who wish you to wake according to their own private rules may be discouraged from entry to your place of sleep.

Dogs may also wake people at any time of day or night, but their reasons to arouse you often involve urgent calls of nature, similar to what occurs to human males with burgeoning prostates. Fortunately, dogs are more trainable than cats. Specially designed doors that allow dogs (and other pets) access to places to relieve themselves are just one of the physical design or behavioral guides that can be used to make sure your dogs allow you to get enough sleep.

Children may also act as inconvenient wake alarms. This situation is common when children are four or five years old and are going through the "monster" phase of nighttime sleep. These monster nightmares are recurrent, frequent, and really, really scary. Though sometimes you have to fully wake yourself up and walk over to the clothes closet to demonstrate that the monsters are truly gone, over time children should be able to sleep on their own without waking others. Since kids need a lot more sleep than adults, the problem, much of the time, is soluble.

Where to Sleep and Wake Up

Sleep is best accomplished in a comfortable, cool, dark place, on a mattress or futon you personally find comfortable and with relatively little ambient noise. Though a major feature of twenty-first-century life, noise wakes people up. Heat, even temperatures above 75 degrees Fahrenheit (24 degrees Celsius), can also wake people up. Cool temperatures, 60 to 68 degrees F., help many sleep better. Light in particular wakes people up.

We are sensitive to sunlight, which has many health effects, mainly positive. However, light in the middle of the night is to be abjured. Production of melatonin, the hormone of darkness, will cease after just seconds to minutes of nighttime light exposure. For the sake of sleep continuity, you want to keep those melatonin secretions pulsating strongly throughout the night.

If your preferred sleeping place does not have blackout drapes, or if there is too much light seeping through your windows, a simple solution is a night mask (sometimes called an eye mask), a covering for your eyes secured by an elastic band around your head. Night masks are useful for long airplane flights and naps because humans are so sensitive to light. Even with your eyes closed, you can see one-third of a light unit, or lux. Conventional, low-level night-lights produce 50 to 100 lux while the sun, on a bright clear day, puts out 50,000 or even 100,000 lux.

As a side issue, hide all your clocks during the night. Clock-watching is a great way to get people to wake up throughout the night, often at the same hour—just like a clock.

Who to Sleep and Wake Up With

Personal preferences must apply. In Western countries parents usually don't sleep with children past the ages of three or four. In Asia children may sleep in parental beds until near puberty.

Throughout human history it has been common for people to sleep together. Beds were expensive and space was dear. However, recent scientific research shows that people generally sleep better, or at least show greater sleep efficiency, when sleeping in bed alone.

Many spouses cannot sleep alone, deeply missing their partner, someone warm to touch and hold and argue with. The practical British, however, often provide separate beds for married couples.

What to Do upon Waking

Open the drapes. Get sunlight. There's a surprising amount of sunlight even when the sky is overcast.

Sunlight resets our circadian clocks and is a particularly powerful biological time giver around dawn. Dawn's early light has more potent effects on us than light at later hours. For example, morning light improves mood far more than evening light. Morning sunlight can be used to prevent and treat seasonal depression, which in its minor manifestations can affect a quarter to half of people in the Northern Hemisphere during the winter, especially those in the northern parts of the United States.

By resetting your internal clocks, morning light also helps your whole body get in sync. When you are jet-lagged, different organs run on their own separate biological clocks. Morning light can reset your whole system, putting your organs back in sync and setting you in a good rhythmic place for doing the things you want to do the rest of the day.

Once you've opened the drapes, get moving. Stretch. Do yoga or Pilates. Walk around the house. If you can, talk with family members or socialize, as all these activities help reset your inner clocks. If time and environment allow it, get up and walk outside, getting healthy morning light, which may also affect your waistline and weight (see the morning brain-warming program below), or if you prefer, bike or exercise in an indoor gym.

You want to move because we all wake up with a literally cold brain, up to a couple of degrees cooler than it will be later in the day. Exercise, defined as any use of your voluntary muscles, increases blood flow to your brain. Walking and aerobic exercise will help heat you up and increase overall blood flow. For your own and your children's sake and for the sake of family peace, you want to wake up your brain fast.

What to Do If Your Preferred Biological Wake Time Is Impractical

A sleep makeover is a way of redoing your sleep and thereby improving your learning, your alertness, your weight control, and your mood. Despite such advantages, lots of people find innumerable reasons to not even try a sleep makeover.

Some of these reasons are very valid. If your body's preferred wake time is at 7:00 a.m. but your job starts at 6:30, and if by turning up late for your job you will lose it, your health insurance, your house, and perhaps your partner, waking up at 7:00 a.m. is not in the cards.

Similarly, children must often wake up for school start times that are ridiculously early according to our biological clocks, with some girls waking up earlier than boys to prepare cosmetics and clothes. Jobs and children often trump the best efforts to wake at biologically sane times—wake times your body wants and needs to perform well.

What do you do? You compromise. If your job starts at 6:30, you have to get up earlier. Morning physical activity plus light will make it easier to adjust to an unnaturally early wake time. Fortunately, morning light is powerful. Timed along with early exercise, it can be used to reset your biological wake times to make them earlier in the day. Ask an owl. That's how many owls survive the Lark Work World.

Still, don't compromise too much. A sleep makeover gives you the chance to live life differently. You owe yourself the chance to at least *try* something new. Then you'll be able to see whether it makes you feel sharper, clearer, happier, and healthier.

Once you get up out of bed, stay up. Use your natural wake time as the time you wake *each and every day.* Stick to it, whether it's warm or cold, spring or fall, weekend or weekday. If you need extra sleep during the weekend, try to get it during afternoon naps.

To make your life rhythmic, get rhythmic. Fitting you and your life to your body's innate rhythms will give you new ways to rest and relax, concentrate, and think. With time, you should also find new sources of energy.

How long will it take? Go through this thirty-day plan and see. It sometimes takes a while because so many of us have been out of sync for so long.

Remember that our bodies and our lives are naturally rhythmic and, with a little application, musical in their flow. Waking up at your preferred biological time is one of the most potent ways to get those rhythms working for you, letting you do and get done what you really want to do. When you're in sync you'll get more done quickly and effectively, generally in less time, and you'll feel more aware and alert throughout the day.

SLEEP MAKEOVER DAY 2: STICKING TO YOUR PREFERRED WAKE TIME

Repeat day 1.

Get up at the same time, adjusting the alarm and the sleep environment as you need.

Finding the right alarm to wake you is significant. When using a clock radio, try different radio stations to wake you at your new habitual wake hour. You can also try waking up to your cell phone ringtones, varying them every three days until you find ones you really like.

If you live in northern climes and have to wake well before the sun wakes the Earth, consider dawn simulators or bright light boxes. Dawn simulators are specially designed to slowly and gradually increase their light, just like the sun rising on the horizon.

Others prefer waking up to bright light boxes. Lots of people feel down in the winter months. Waking up and quickly turning on a bright light can improve mood and more rapidly wake up your cold brain.

Light boxes are special, really bright lights. Light boxes used for seasonal depression are meant to pump out light that resembles sunlight in intensity and are quite costly. However, they can be used for far more than seasonal depression. Bright light boxes are useful for getting people up who find waking in the morning next to impossible (I use them for narcoleptics). In the evening you turn them on to increase alertness. As you'll see later, if you time light use correctly, it can also help you get to sleep.

Light Devices

There are many different kinds of dawn simulators and light boxes, and they change all the time. Here are a few that patients recommend:

Dawn Simulators

BioBrite Sunrise Clock

SunRise System Dawn Simulator

Apollo Health DayBreak Duo Dawn

Light Boxes

The Philips GoLite P1 (small, cheap) or the Philips GoLite Blu (expensive) both work through bright blue light.

More traditional full-spectrum boxes—those by the Sunbox Company, like the Sunlight Jr.—are also effective.

Morning Brain Warming

As we literally wake up with a cold brain, I would recommend:

Fulfill any required acts of nature.

Now that you are slightly conscious stretch your legs and arms. Yoga poses like downward dog and child's pose are excellent here, but putting out your arms and twisting right, then left, also works well. Stand up, and then stretch your hands toward your toes.

Put on old clothes.

Walk for at least five to ten minutes, briskly. *Do not* look at your e-mail first! Leave your cell phone alone! If at all possible, walk outside. If you are living in northern Saskatchewan in the middle of winter, find a nice pleasant hall to walk in.

Stretch again for one to two minutes.

Shower.

If you must, glance at your computer and cell phone, recognizing you are now awake enough to deal with the urgent messages you may find there.

One advantage of morning brain warming may be better weight control. Studies done by Colin Shapiro at the University of Toronto looked at older folks trying to lose weight. Some rode their exercise bike in normal, artificial light, others in bright light. Though both groups lost weight, the bright light exercisers lost fat mass. They seemed to get more muscular.

Shapiro did not know why. New data in animals (University of Nottingham, 2009) shows that brown fat is activated by sunlight. Brown fat is common in hibernating animals, putting out enough heat to keep them warm in winter. Though humans may not have much brown fat, it is vastly more efficient at using up calories than normal white fat and negatively correlates with weight. If daylight provokes brown fat activity, morning walks in sunlight might help us lose weight and waistline.

So when you wake up to your alarm, really wake up. Get up from the bed. Once you're up, you're up. Stretch those stiff muscles. If you can, get some clothes on and take a walk outside, or indoors in bright light, getting light and activity that will warm your brain and clear your head. Wake time is the time to begin, to recognize the living wonder of the day, and to start it.

For owls, working a regular daytime job means becoming a shift worker. Shift workers often define their jobs by leaving out the fourth letter of the word *shift*. Even if you're not paid to wake up early, as happens to owl college students trying to attend and then comprehend an 8:00 a.m. lecture, things can get tough for owls.

Fortunately, owls have tools they can use. First is light. Sunlight in the morning makes owls more like larks.

Next comes exercise. Exercise also shifts your internal morning clock earlier.

Combine light with exercise, as in morning brain warming with walks outside, and you can make even severe owls larklike. Some of my owl patients found they could do a whole series of new jobs once they started walking in the morning. For many, thirty minutes marching around the neighborhood or striding on the beach got them up and ready.

And they felt good. Light improves mood. So does exercise. In some European countries, outdoor morning walks are used to treat all kinds of clinical depression.

So wake your brain with a morning walk. It should make you and your brain happier and will give you time to think out what you want to do with the day. The insights that come with REM sleep pop into awareness more often when people take a morning walk.

SLEEP MAKEOVER DAY 3: GOING TO SLEEP AT YOUR NATURAL, PREFERRED SLEEP TIME

You've been waking up at the same hour now for two days. Now you're ready to fall asleep at your preferred sleep time.

One way to make the transition to a different wake time easier is to do something relaxing right *before* you go to sleep. The many different techniques for physical, mental, and social rest you'll soon learn will also make the transition to sleep easier.

You can start with one version of the first physical rest technique, described more fully in the next chapter. Sitting in a comfortable chair, breathe in abdominally to the count of four. Then breathe out to the count of eight. Feel your belly move up and down as you breathe.

Remember, you need to give yourself enough sleep time to eventually feel truly rested. Yet too many of us feel wired at sleep time. It's not easy moving from the constant motion of wakefulness to the internal calm required for us to fall asleep. But getting in sync produces lots of benefits, from greater work productivity and social ease to feeling much more alert and alive throughout the day.

Sleep Time Practicalities

If you have a job, like mine, where you are on call twenty-four hours a day, think about making it hard for someone to wake you up without a reason. If you can, turn off the phone's ringer. If you can't, record an outgoing greeting on your voicemail that politely makes it clear you're not available—unless something important comes up. With many cell phones, you can set a special ringtone for emergencies. Remember, you need food to live. You also need sleep to live, and to function.

If pets are not habituated to your going to sleep at a set time, talk to them and cajole them. Many people tell me they feel at least as comfortable sleeping with a pet as with a spouse. Recognize on beginning to sleep with a pet that their different biological clocks may take quite a while to fully adapt to yours.

Further Practicalities: Bed Partners

Your bed partner may be easier to remonstrate with than your animal companions but may still want you to go to bed at his or her very different sleep time.

Politely resist—at least during your seven-day sleep makeover.

Many bed partners want to go to bed at the same hour and get up at the same hour as you do. Some people find they can make such biological clock adjustments very easily. As I have said, the ability to shift inner biological clocks is itself a genetically based trait, so you may be able to go only so far in your negotiations. Some people may need to go it alone for a week in order to fully experience their sleep makeover.

Recognize that getting to bed at a time outside your biologically preferred sleep time can become like a chronic version of jet lag. Unwittingly or unwillingly, you may have been doing this for many years. Instead, consider what your particular human design tells you that you need to do in order to get fully restful sleep.

Still, many people tell me they don't want to go to bed at different times than their spouses no matter what: "If I get into the bed later, it'll wake him up. He doesn't like that. He really doesn't." Others protest that going to bed at different times denies them whatever opportunity they have for sex.

Hearing that story makes me sad. Sex is (on most occasions) an excellent form of social rest. Yet Americans talk about sex, dream about sex, and fantasize about sex a lot more than they engage in it.

Do you want to experience sex only when you're exhausted and about to go to sleep? Studies done on couples in sleep laboratories—admittedly not the most romantic of settings—show that sleep does *not* improve after sex, despite the claims of Woody Allen. There are many times during the waking day when sex can be scheduled, times when the fact that you're feeling alert and alive becomes a major plus. One of the many benefits of good sleep is that it makes you alert and aware for many other pleasurable and varied activities.

Do people have sex, then watch their partner fall asleep while they themselves feel really, really awake? Happens all the time. Males often experience *le petit mort* and start snoozing right after climax, exactly when their female partners feel more aroused. This is another great reason why sex should not always take place right before sleep.

Will going to bed at different times inevitably wake your bed partner? No, especially if your sleep hours become habitual and regular.

The truth is that people frequently wake up during sleep all through the night—except they don't remember waking. The biological line between being asleep and being awake can be very fuzzy. Generally, you have to be awake about six or eight minutes or more before many folks recognize that they are awake at all.

That's because sleep causes amnesia. When you're asleep, you don't remember what you did during sleep. Even perfect sleepers wake up fifteen to twenty times a night, usually for a few seconds to minutes. And no, they don't remember those short awakenings.

My sleep laboratory's waking-up record is held by a woman who walked out thanking the technicians for her best night of sleep in four years. Looking at her brain EEG, we saw that she had awakened for at least a few seconds over 1,200 times.

Normally if people wake up that much, even for brief periods, they feel like they hardly slept at all. Not this woman. That she had experienced more than a thousand brief awakenings was totally unknown to her. She never slept more than ninety seconds without some kind of awakening. We all developed eye strain looking at the interrupted wave patterns defining awakenings on her sleep EEG, which was chock-full of apneas (stopped breathing episodes) and sleep-stopping leg kicks. Of these events she had no recollection.

The converse of this story is also true—people who are asleep often think they're entirely awake. One study done at Henry Ford Hospital in Detroit observed people in light, stage 1 sleep for ten full minutes. Brought back to consciousness, they were asked to describe what had happened to them.

Fully half of them thought they were awake the whole time. People were shown their sleep records, with nice illustrations of when they went to sleep and were woken up. They were then shown the videotapes. Some still did not believe they had been asleep.

In my own experience, in sleep labs and outside them, people may be sleeping and loudly snoring right in front of you. They'll still tell you they were awake the whole time.

So if you say to your partner, "You were asleep," and your partner replies, "I was awake!" don't argue very long. In our modern world people are so sleep deprived they frequently experience microsleeps of a few seconds' to minutes' duration that they do not recognize. These microsleeps happen everywhere—driving the car, at work, at school, in the movie theater, dining, even during sex. As Torbjorn Akerstedt of the Karolinska Institute and others have shown, microsleeps often occur among night-working train conductors and airline pilots. That's something to consider the next time you take a fifteen-hour flight across the Pacific. It's also why there are two or even three teams of pilots on long flights, sleeping in concealed cabins few passengers notice.

SLEEP MAKEOVER DAY 4: DEVELOPING A SLEEP RITUAL

Preparing for sleep can be as important as what time you go to sleep. Your body is not built to dart suddenly from extreme mental and physical activity into immediate slumber.

Rest and activity demand transitions. With the exception of professional athletes, most of us can't suddenly downshift from high physical activity into immediate sleep. Neither can most of us worry intensely about jobs, family, kids, and health insurance and suddenly and pleasantly fall into a restful sleep.

That's why you need a sleep ritual. Your sleep ritual helps you relax and downshift in the hour before your preferred sleep time.

Do you floss your teeth before bed? Brush your teeth? Congratulations, you've already got elements of a sleep ritual percolating.

It sounds odd, but in order to fall sleep you first have to *feel sleepy*—or at least relaxed. The brain needs to be properly set up to make those amazing transformations required to fall asleep, which then prepare you for all that wonderful rewiring and restoration that sleep produces. So though there are some general rules to make sleep rituals effective, you will need to make your sleep ritual truly *yours*, a time and a set of small, pleasant actions that fit your needs and personality. To sleep well, you need to get your brain and body ready.

Before you start creating your own personal sleep ritual, here are the answers to some questions my patients ask:

Why is it called a sleep ritual?

Because everything works better if much of your presleep period becomes routine, rhythmic, and ritualized. As you perform a sleep ritual night after night, your brain and body get used to the idea that this little group of behaviors will help you sleep and will provide you complete rest. If you have to describe it in a single word, a sleep ritual is conditioning. Without thinking, you can peaceably and pleasantly go to sleep.

What sorts of activities do people do during a sleep ritual?

Simple, pleasant, ordinary, relaxing things. In weeks 2 and 3 of this plan you'll learn a lot of different ways to rest, some of which you may want to add on to your personal sleep ritual.

Little activities work well. You floss and brush your teeth. Turn down the bedcovers. Put out the clothes you will wear tomorrow, whether for work, casual wear, or exercise. Listen to music. Read. Reading can be very useful for helping you to fall asleep (you need not read this book before you sleep, but it is permitted).

What if some nights I don't have enough time to do a full sleep ritual?

Do what you can. Often the simplest, quickest elements of a sleep ritual still provide guidance to the brain that this is indeed sleep time. If you couple conditioning with the powerful influences of your internal biological clocks, which will increasingly synchronize as you regularly go to bed at your biologically preferred sleep time, things really get going. As you get synced you begin to fall asleep quickly and comfortably. With time, most of us will fall asleep without any effort.

Which is how things should be.

Too many people would rather do lots of things other than sleep. To them, going to sleep is a job. In my experience, if you turn sleep into work, you probably won't sleep very well.

Sleep should be as natural as breathing. Like food, it's something to look forward to, and most of us get to do it every night. A sleep ritual engages the pleasures of relaxation prior to the amazing body transformations performed by sleep.

So you've turned down your bed, cleaned your teeth, put out the clothes, kissed your partner good night. Now what?

Reading as Part of a Sleep Ritual

Reading is good for the brain. You get mental stimulation through going into the world of the book, whether it's this world or some other planet or galaxy. You touch another's consciousness, hear that person's thoughts and respond to them, wondering, calculating, analyzing, identifying, and musing. Movies can be great but are too arousing for many people to help them sleep. But as part of your sleep ritual, reading can help.

True, many books are far too arousing to aid your presleep rest. Mysteries and thrillers, which are constructed to keep you turning the page, may keep you awake, and comic books and graphic novels likewise can become too gripping to be put down.

For your sleep ritual you want to read something that will literally let you fall asleep. The standard advice is: "Read a book you should have read in high school but didn't."

Classics are the glory of our literature, but they have not always been popular. Not everyone likes reading the *Iliad* before bed, though there are a few strange people, like me, who do.

From long experience I can recommend at least seven different categories of books to keep nearby for reading before you go to sleep.

History generally places your own life in perspective. History books help you see the long view, to understand where we all are now. Historical knowledge, combined with elegant syntax, is often a great aid in making you relaxed enough to fully prepare for sleep.

Art history books often combine beautiful pictures with learned commentary, letting you pass easily from reading to viewing to reading. Seeing and reading about beautiful things often provokes changes in dream content and creates images throughout your mind that will relax you and make it easier to sleep. There's something very restful in pleasantly touching the visual cortex.

Poetry is often dense, with more than one mental image operating simultaneously. Poetry is also usually rhythmic, which can also aid falling asleep. Take, for example, Longfellow's epic poem, *Song of Hiawatha*. Longfellow wrote *Song of Hiawatha* in the same rhythmic meter as the great Finnish classic, the *Kalevala*. It possesses a poetic rhythm that many English speakers find potently sleep inducing. Fortunately, extraordinary poems abound in our language. Some patients tell me that simply reading prefaces to the collections of seventeenth-century English poetry that I recommended sent them into immediate slumber.

Travel books should and will take many of us far away. The visuals they create often make the transition to dream sleep more enticing.

Historical fiction can work in a similar way to history texts as a foundation for your sleep ritual, yet it often possesses greater sweep and narrative imagination. Living other lives during long-ago times frequently proves enchanting.

Biographies, in a similar way, let us know lives other than our own in a manner that instructs and entertains. Good biographies can make us feel in touch with people we deeply admire, people we would have loved to meet and know.

Self-help health books can be useful prior to sleep, especially when they include chapters about sleep. There are some people for whom nothing induces sleep like reading about sleep.

You may indeed already possess this full variety of books on your shelves, ready and waiting to be used. If you don't, libraries are filled with books that should help you relax and easily fall into sleep. For purposes of going to sleep it's best to read on a chair, not in the bed itself. If you read in bed, your brain gets used to the idea that you can simply stay awake there. That's why almost all experts on insomnia advise reserving the bed for sleep and sex only. So if you can, read in a chair next to the bed, or in another room, as part of your sleep ritual. And it's a good idea to have volumes of several different categories scattered around in case you find the book you're reading too engrossing.

Following a time of reading, it often helps to do a brief cycle of deep breathing or simple yoga. Such maneuvers can make you feel profoundly relaxed. Lots of people tell me it helps them to start deep breathing (physical rest technique 1) as soon as they fall into bed. They breathe in to the count of four, out to the count of eight, expelling the air trapped at the base of the lungs. As they breathe in they feel their belly soften and rise, which sets them up for other rest techniques we will cover in the next few chapters.

TV or Not TV

Don't let anyone fool you—commercial TV wants you awake and alive. All night, if possible.

Television is remarkably lucrative, and TV executives spend well-remunerated lives discovering new approaches to entice people to lock their eyeballs onto the tube. Their industry and insight have been crowned with success. Nielsen, the prominent rater and recorder of television use in the United States, has reported that televisions are on for eight hours and eighteen minutes each day. Americans sleep on average fewer than seven hours. No wonder people usually mention TV first when asked how they try to rest.

Yet is TV truly restful? Think about those programs you watch late at night. Don't forget, TV programmers need your eyeballs open to sell you advertising. Programming at night often involves quick shifts in loudness. Nighttime commercials are often broadcast at high decibel levels, considerably higher than during the daytime. Nighttime visuals also may run riot. Diagonals, jump cuts, and fast montage accompanied by jazzy, intrusive music work together to keep your fingers off the off switch.

Nonetheless, many people feel that TV puts them to sleep. They watch the news, discovering who was murdered and which daytime TV teen stars are divorcing, followed by deeply intoned predictions of titanic hurricanes and deadly tornados. After all that, they expect to promptly nod off.

My mother is one of these people. She falls asleep to the TV. I believe she falls asleep when she does because her body clocks have decreed that it's her sleep time, but try telling her that. For some folks, it does indeed appear that TV operates as its own sleep-inducing ritual, part of a routine that aids our fall into slumber. However, for adolescents, nighttime TV appears to correlate far more with behavioral problems, particularly depression. A study done at the University of Pittsburgh and Harvard in 2009 showed that the more TV teenagers watched, the greater their risk of depression during the next seven years. More TV also generally means less sleep, and less sleep means worse cognitive function and lower grades. Among adults, late-night TV use is a prominent feature in insomniacs, many of whom eventually become depressed. Another unfortunate biological effect of TV comes from the light emitted by the TV itself.

While morning light sets our internal clocks earlier, light in the evening sets our biological clocks later. Television sets put out a lot of light. The effect in many people is to make them stay up later, cutting down on their sleep time and making it harder for them to wake up in the morning.

If TV puts you to sleep, okay. But if you're frequently waking up throughout the night, reconsider. And if you find yourself staying up to watch a favorite show, think about recording it. You can watch it later, when you're alert and really able to enjoy it.

So try out these elements of a sleep ritual one by one. See what works for you. With time, they usually become seamless. Sleep is then no longer a project to accomplish or a fond wish but an effortless fall into powerful, restorative rest.

SLEEP MAKEOVER DAY 5: LEARNING NOT TO WORRY

You are now getting in the habit of going to bed and getting up at your preferred biological times. This regularity helps to sync your body and your mind. You are also working on improving your sleep ritual every night, subtracting or adding little behaviors that work for you.

But if you want to sleep you may need to teach yourself something really important—learning not to worry. Training yourself not to worry can help you in the rest of your life as well as with getting a good night's rest.

Plenty of things stress us out. Some are big, many are small, but put them all together and they can really get to us. They often get to us most at night. When the day is over and all its distractions gone, we have time to reflect. At times of economic crisis, that reflection can lead to endless worry.

Worrying can prevent sleep, but worrying about sleep is extremely effective in preventing sleep. That is why brief techniques of cognitive therapy can help. Cognitive therapy trains the brain to think in terms of solutions, not problems.

The founder of cognitive therapy was a Philadelphia psychoanalyst named Aaron Beck. Back in the 1960s, Beck noticed that not only were his patients not being helped by classical psychoanalysis, but quite a few of them were getting worse. In particular, many of his depressed patients became significantly more depressed while undergoing psychoanalytic treatment. Beck came up with cognitive therapy because he wanted something that would work.

It does. Roughly put, cognitive therapy is rationalization writ large.

A quick and dirty but efficient way of beginning cognitive therapy is to perform worry time—to write down the five or so major things that are really bothering you. But—this is important!—do this *hours* before you go to sleep. Immediately after you write down your worries, you do something equally important: you write down their solutions.

A good time to perform this simple version of cognitive therapy is right after the evening meal. You've eaten a nourishing, well-balanced feast, filled with lots of plant-based foods whose fiber may aid your sleep. (There is some evidence that eating lots of protein in the evening will keep you up, but it's not strong. Eating late at night, however, usually does keep people up.) You're feeling comfortably full, and since it is evening and your body core temperature is rising, you should also feel pretty alert.

To start cognitive therapy, find a small notebook. If you prefer, you can write on your computer or PDA in a secure file. Next, allot four or five minutes for your cognitive therapy homework every evening.

First, you write down four or five things that are really bugging you. Maybe it's the loss of a job. Difficult-to-treat back pain. Worries about paying for your kid's school. Write these problems down in a few brief words.

Then, write down their solutions—what you're presently doing or plan to do to try and fix those problems.

You point out all the different ways you're trying to find that job, from networking to job lists you've created to leads your best friend from high school gave you last night on the phone. You describe the different treatments you've tried for back pain, listing those that helped and what you're trying now, perhaps including the different types of stretches and yoga routines you've just learned. You note the time and place you're going to talk to your kid about how much college costs and what her options include, ranging from student loans to work-study to matriculating at a different school.

Some problems just don't have solutions. Some diagnoses cannot be effectively treated, just as some political and social facts of life cannot be changed. But even if we cannot change fate we can do as the Stoics taught and change our response to fate. Usually even in the worst situations there are things we can do to function better, to engage others to help us, to consider other forms of action.

Cognitive therapy helps because it gets your brain to fixate not on problems but on *solutions*. You have to move forward and get things done. Training your brain to constantly think about creating new solutions itself produces real benefits, usually including a better, more optimistic mood.

Another advantage of doing cognitive therapy, especially right before or after the evening meal, is that you don't have to think about all those daunting problems right before you go to sleep or when you wake up. Cognitive homework helps you let go of problem thoughts when they creep unbidden into consciousness. Ah, you tell yourself, you've already done the work. You've written down your solutions. You have a plan to handle things. There's absolutely no need to ruminate about all that stuff before sleep or in the early morning hours.

The beauty of using a notebook or computer for cognitive worry time is that you then get to watch your problems, and solutions, change over time. Not infrequently, problems go away. If they do not, you can check solutions you tried in the past, evaluate whether they helped, and then plan new actions to fix what's wrong.

Cognitive therapy is about far more than creating a positive attitude. It's about coming up with solutions and learning to get better and better at creating new solutions.

Cognitive homework does not require much time. Some people do their writing for a half hour every evening. However, even a few minutes can push the process a surprisingly long way.

You can also use cognitive worry time to tackle issues specific to sleep. Couldn't sleep last night? Write down what you believe are the reasons. Next, write down what you plan to do to make sure your next night will improve.

With time, you will see that you can use cognitive therapy to provide solutions to many aspects of your life, ranging from economic to social to psychological issues. And your cognitive solutions tend to improve as your capacity for sleep and rest improves. It's a welcome side effect that comes from paying attention to the power of rest.

SLEEP MAKEOVER DAY 6: A HOT BATH BEFORE BED

To get your brain to shift from wakefulness to sleep is not easy. Many brain structures and systems have to move simultaneously from one state to another. Frequently they don't. Often people jerk their muscles as they are falling asleep, a "sleep startle reaction" that can be witnessed at many concerts and movies or, more embarrassingly, at business meetings.

Fortunately, there are direct physical ways to hasten the transition to sleep. No, I don't mean sleeping pills. There's a simpler, more natural way to relax and promote sleep—a hot bath.

Lots of researchers think one way the body controls alertness versus sleepiness is through an inner temperature "sleep gate." Open that gate and you fall asleep.

This sleep gate opens and shuts through changes in body core temperature. Body core temperature is the temperature inside you—for example, in your spinal cord. A rectal thermometer is a reasonable proxy for body core temperature (not an oral thermometer, I'm afraid). When body core temperature is rising, we grow more alert, as normally occurs in the evening. If body core temperature goes down, we get sleepy. If you can get your body core temperature to rapidly decrease, this seems to open the sleep gate, signaling the brain and body to fall asleep—and remain asleep.

Janet Mullington and other researchers tried hot bath therapies at different biological clock intervals. They found that the closer the hot bath was to people's preferred sleep time, the quicker people fell asleep. Their sleep was also more continuous, without as many interruptions. That is important. Though people usually can't recall them, brief awakenings, especially if there are more than the usual fifteen to twenty that happen with good sleepers, can make people feel tired and unrested the following day. Hot baths before sleep also increased people's deep and REM sleep. A hot bath is an efficient, cheap, and natural way to promote better sleep throughout the night.

Temperature is the key. To get your hot bath to really work for you, you need to sweat. It helps to surround your spinal cord in water that feels quite hot. Mullington's group spent a half hour in the bath, but many will get improved sleep from sweats of just two to three minutes. Sweating usually means you've gotten your core body temperature up a degree to a degree and a half. Almost immediately after, your body cools down. That quick cooling may be what pries open the sleep gate.

Prefer hot showers? Sorry, but showers rarely raise body core temperature enough to help us sleep. The behavioral stimulus of a shower at night is soothing to some, though others use the stimulus of pulsating water to wake up in the morning.

Aromatherapy

Many people use aromatherapy to help them sleep. My personal experiences here, however, have not been great. Some of my patients have developed allergies though others love aromatherapy. Certainly the behavioral conditioning of lying in a tub with lovely scents wafting across the water helps many fall sleep.

Add Hot Baths to Your Sleep Ritual

Sleep rituals help us sleep better through conditioning, and it's easy to add a hot bath to make your ritual more effective. So try this: following the evening meal, do your cognitive homework. Next, starting an hour before your preferred sleep time, begin your regular sleep ritual. Do it every night. To improve things yet more, take a comfortable hot bath right before you intend to go to sleep.

Make the water hot but never painful to your touch. Test the water before you get in, making sure it feels properly warm, churning the water lightly with your hand. If you want to make the whole experience more restful, turn the lights off before entering the tub.

Light is stimulatory. It helps keep us awake, and it restores alertness if turned on during the night. Still, when you bathe, please do not make your bathroom totally dark, as you will need to see what's going on. Small night lamps, candles, or lights in neighboring rooms all can provide enough light for vision. You goal is to make the room feel warm and inviting. Bathing in relative darkness can help begin the process of making you feel very relaxed. You might also want to have relatively low lighting in your bedroom so that when you return from your bath you won't be jolted by bright light and will be able to continue appreciating the peacefulness created during your bath.

Deep breathing also promotes relaxation. Enter your hot bath when it's a third to half full. As the water pours over you, rushing over your feet and between your legs, start to breathe deeply. Feel your abdomen rise and fall. Breathe in and out slowly.

As time goes on and the water rises, you'll feel the tug of the water's surface pressure on your belly. Sense the water spreading over your belly and chest.

Now breathe easily and slowly, in and out. Try to breathe out for about twice as long as you breathe in. Yes, make those exhalations last. Sometimes it's easier to start the process by counting out your breaths. If counting to four is too long, start by breathing in to the count of two, out to the count of four. Feel the water's surface tension as your belly and chest go up and down, rising and falling.

Try to think of pleasant images as you slowly breathe. You might walk across the soft, pliable sand of a glittering black beach. You might recall a favorite hike you took through the mountains and countryside. Breathe in and breathe out, trying to put your breathing in rhythm with the pace of your imaginary walk. Feel the muscles at the back of your neck relax as you exhale.

Before you know it, you may begin to sweat. There's no need to sweat for very long. Feel the drops tickle your earlobes, or sense a slow rill of water flow down your forehead. After you've sweated for two to three minutes, you can leave the bath, though some will experience more deep sleep with longer sweat periods.

Once you towel off, you should feel more relaxed. Perhaps you already feel sleepy. If you like, you can immediately enter your bed following the bath. Others prefer to read a little bit. Remember to time your sleep as close as you can to your preferred biological sleep time.

What Hot Baths Do

Hot baths can be used as a type of physical rest at any time of day. But at night, hot baths achieve special, multiple effects.

Deep sleep time is increased by hot baths. The more deep sleep you have, the more growth hormone you produce. Growth hormone grows muscles, joints, and brain cells.

Hot baths create better sleep continuity. With fewer awakenings from sleep, you will experience greater alertness throughout the following day. More deep sleep and REM may also translate into better memory and learning.

Hot baths do not work for everyone. For some of us they are simply too stimulating. If you've had frequent urinary tract infections, some experts believe hot baths will make them worse. Yet baths have been popular with people for thousands of years for very good reasons. Baths can help you relax, rest, and sleep.

When you wake the next morning at your biological preferred time, don't be surprised if you feel more refreshed. Neither should you be surprised if you went to the bathroom less frequently during the night. Better sleep continuity means your bladder reflexes are turned off longer and more consistently. Whenever you wake up during the night, those reflexes get turned back on, making you want to urinate. As hot baths promote sleep continuity, less waking means fewer trips to the bathroom.

Natural things work. Nothing is more natural than rest.

SLEEP MAKEOVER DAY 7: AN EVENING WALK

You've done a lot in the last six days. You're going to bed at your biologically preferred time and waking up at another. You're starting to do brief cognitive techniques that get your brain thinking of solutions rather than problems. You have a sleep ritual going, behaviorally conditioning yourself to easily fall asleep, aided by the fine, relaxing measure of a hot bath.

Next it's time to do something else perfectly natural—take a walk after your evening meal.

Years ago, Jim Horne at Loughborough University did a series of experiments studying exercise and sleep. People who exercised about three to six hours prior to sleep slept better. No exercise, no improved sleep.

Temperature and sweating mattered. If Horne put his exercising students through a cold shower, the sleep-inducing effect disappeared, one of the early demonstrations of the sleep temperature gate.

You don't want to exercise right before sleep unless you suffer from restless legs or other sleep maladies. If you exercise right before sleep, especially if you exercise intensely, you'll probably keep yourself awake longer. I found that out decades ago while teaching at the University of Texas–Houston and playing on an Ultimate Frisbee team that practiced at Rice University (I was both the oldest and worst player). When we did our team practices at night, going till 9:30, I couldn't fall asleep for hours.

Timing exercise right can promote sleep. Exercise three to six hours before sleep and your sleep should improve. As you'll soon see, exercise is also mentally restful.

Trying Out the Walking Cure

You've spent your four or five minutes writing out your cognitive worry time in the early evening. Your next step is to take a walk.

During that walk you can think about all that you've written. What solution that you tried is working? Perhaps another attempted solution is not getting you anywhere. Is there something else you can do to try and fix that problem? Your thoughts should sharpen as you walk.

It's also good to walk with someone. Walking with a friend or partner promotes social rest, as you talk, joke, and help each other figure out how to combat the stresses of life.

Another advantage to an evening walk is slimming the waistline. Walking after meals slows digestion. That slows glucose uptake, which in turn slows insulin release. In the last part of this thirty-day plan you'll discover how nicely such walks may help you control belly fat and weight.

If it's spring, summer, or fall your evening walk may coincide with the last rays of the evening sun. Treatment trials have been done using bright light therapy to improve insomnia. Using bright light

Things Not to Do Before Sleep

Caffeine after midday. The caffeine in coffee, tea, or energy drinks can have effects that last twelve to sixteen hours after you drink. If you're having trouble sleeping, leave caffeine to the morning. There are documented cases of young women with diagnosed narcolepsy whose problems disappeared after they cut out their two cups of morning coffee.

Alcohol as a nightcap. Alcohol is used by perhaps 5 percent of the U.S. population to fall asleep. In people who don't normally drink much, alcohol right before sleep will lead to fifteen to twenty-five added awakenings each night. Also, the worst insomniacs I've seen are either alcoholics or former alcoholics, due to the profound effects of alcohol on the brain. If you like to drink, drink in the evening hours before you plan to sleep, a time when alcohol's positive cardiovascular effects are strong and negative effects are minimized.

Eat a lot. Many suffer from night eating syndrome, waking up to a whole meal around midnight, often after

boxes in the evening, several hours before sleep, added a full hour to insomniacs' sleep. Evening light exposure, as long as it takes place several hours before sleep, can help you sleep better.

If you're living in the north and there's a blizzard, you may not be able to walk outside every evening. Walking on a treadmill also works. Do what you can. You can certainly walk or stroll *around* your home. It's also fun to have a conversation, whether you're clearing and cleaning dishes or pacing the living room chatting on a phone. Talking on the phone with parents or children or friends does not mean you can't move. Evening physical activity helps most people relax. It also helps you sleep.

having eaten a meal earlier in the evening. Eating a lot at night promotes more wakefulness and lots of weight gain. A glass of milk before sleep or a small, complex carbohydrate snack is probably okay, especially for diabetics. Lots of sugar at night may cause more awakenings, however.

Writing out plans for the next day. It's generally better to do planning, as well as worrying and cognitive therapy, hours before you go to sleep. Otherwise, the thoughts may keep you up.

TV right before sleep. Yes, this is controversial, as many use TV as a behavioral cue to fall asleep. But as we saw, visual and sound effects are often used in late-night TV to keep you watching and awake. Also, the effect of light from the TV may keep people up later and longer by shifting inner biological clocks later.

Sleeping pills. Though they're very helpful for treating jet lag or for temporary periods of sleeplessness, sleeping pills are habituating, as the brain gets conditioned to needing a pill in order to sleep. Even the newer "nonbenzodiazepine" sleeping pills can impair memory and performance.

Writing, moving, socializing, if done at the right times and in the right ways, can all help you sleep. Sleep then becomes a pleasure, something to look forward to every night. You learn more, become more creative, and help improve your health. Good sleep also aids your ability to rest in other ways, mentally, socially, and spiritually. Your sleep makeover then produces surprisingly powerful results that build and build over time.

Summary

In school there are the three Rs, reading, 'riting, and 'rithmetic. Your body also has three Rs—rest, restoration, and renewal.

From what you've read, you can see that sleep is a part of rest and is as necessary to your survival as food. We've learned that good sleep does many things. It

helps control weight

renews memory and improves learning

helps prevent depression and improves mood

improves complex thinking and makes new brain cells

fights off infection

prevents heart attacks

strengthens and maintains normal biological clocks

Sleep does all this through literally rebuilding and rebalancing the cells of your body and their interactions. Without sleep you don't make new brain cells. Without sleep you don't lay down memories. In sleep, you produce growth hormone, recontouring and refiguring your body.

To get all this wonderful stuff, you have to sleep right. Now that you've done this chapter you probably know

how much sleep you really need

when you need it

simple, easy tools to get it

These tools include cognitive homework, hot baths, properly timed work, creative sleep rituals, properly timed walking, and good times to get light.

It's important to follow these rules of sleep during your thirty-day rest program *and* beyond. It takes time to get used to sleeping right. But once you see how easy it is, you will probably be set for good sleep for the rest of your life. Hopefully, you will have a more rested, zestful, balanced life, which gives you a much better chance to stay healthy and feel whole.

Sleep is one kind of passive rest. Now it's time to learn how to actively rest in the daytime or the night, wherever you are.

chapter 3

PHYSICAL REST

Physical rest is an active form of rest, a conscious, directed way in which you use your body's basic processes, like breathing, to calm and restore body and mind. Physical rest makes both body and mind more relaxed and makes you more able to concentrate. These techniques are easy to learn and are right at your fingertips. The better you become at these techniques—and they improve with practice—the more easily you will deal with the stresses in your daily life.

You may be sleeping better already. Now, in day 8 of this thirty-day plan, you will start using simple active rest techniques that keep you conscious, focused, and attentive to what's going on around you, plus let you have some fun doing the most ordinary things. But first, why don't you try a brief quiz? Pick the best answer(s) for each question:

1. When I sit in front of the TV, I am
a. passively resting.
b. actively resting.

2. The best place to physically rest is
a. at home.
b. at work.
c. on vacation.
d. anywhere.

3. If I rest well, I should feel
a. relaxed.
b. calm.
c. centered.
d. more alert.
e. all of the above.

4. To properly physically rest, I need to be
a. physically fit.
b. fully alert.
c. tired and needing to rest.
d. sleepy.
e. flat-out exhausted.
f. none of the above.

5. The best time to rest is
a. when I'm bored.
b. in the late morning, before lunch.
c. in the evening.
d. just before sleep.
e. anytime I want.

6. The requirements for physical rest include
a. the ability to breathe.
b. the ability to focus.
c. the capacity to think deeply.
d. help from friends.
e. a sense of purpose.
f. a plus b.

7. The best age to learn how to physically rest is
a. adolescence.
b. early adulthood.
c. prime maturity.
d. just before retirement.
e. in the next few days.

8. I can learn how to physically rest from

a. myself.

b. this book.

c. my friends.

d. a plus b.

e. all of the above.

Answers: 1. a; 2. d; 3. e; 4. f; 5. e; 6. f; 7. e; 8. e.

Rest Is Active

There's a big difference between actively resting and passively resting. Knowing the difference can make life easier.

If you're like a lot of people, you may be worried these days. Perhaps you're sitting in your office chair staring at the computer screen. You may not like what you're seeing. Perhaps your credit card bills are too high. Or it's just today's news: there are more layoffs in the industry where your brother-in-law works, plus a new storm gathering in the west. The news is no worse than usual, but it sets off your mind in a thousand different directions. Your thoughts keep searching and changing, moving you're not sure where.

Are you resting now? In a sense. You're probably passively resting.

I say "probably" because *passive rest* usually means your body is still, which at the moment may not be the case. As you look over that screen you're probably not physically quiet, as your legs twist under the desktop, your toes tap in your shoes, your index finger stabs at the mouse then folds back and curls upward to massage the back of your neck.

Passive rest does provide a time and a means to restructure parts of your body and brain. As you walked to your desk, you stepped wrong on your left foot as you sat down and minutely twisted your ankle. Now that your finger and toe tapping are finished, passive rest gives the cells in your ankle the chance to check out the damage to your synovial joint. Broken structural proteins are cut up and taken away as the body begins the process of inflammation that initiates body rebuilding. Inflammation is necessary to survival, but it can go too far. If too prolonged and sustained, it can provoke chronic diseases like cancer and atherosclerosis.

But right now inflammation is doing its proper job. The damaged structural foot proteins are removed. New fibrils start to sprout. Messages from arteries and nerves are quickly sent to heart and brain, making you shift position as you sit. There's a constant flow of information going forth now to your sympathetic and parasympathetic nervous system, your brain, your liver and muscle cells. This extraordinary activity takes place almost entirely below your conscious awareness.

Yes, you are *at rest*. But you are not actively *resting*.

Active rest is conscious. Active rest is under your control. Active rest is goal oriented and directed. With the techniques of active rest you are about to learn (and there are many more), you will rest your body and mind at will, at your leisure—and for your pleasure. Doing active rest you will be learning the whole time, restoring your body and developing parts of your brain that will make future techniques of rest easier to learn and use.

Physical rest is one of the many forms of active rest, which include mental, social, and spiritual rest. There are many ways to achieve physical rest. Perhaps the simplest is just to breathe.

DAY 8: PHYSICAL REST TECHNIQUE 1— DEEP BREATHING

To many mystics, the world begins and ends with a breath. Inside that single action starts life and the known universe.

Our goals in resting are more modest. We want to learn to rest anywhere, anytime, in ways that restore us, calm us, relax us, and make us alert. That means we have to learn how to breathe.

Breathing is a completely natural action, but when we do it well, breathing is a little different from the unthinking process that nor-

mally keeps us alive. The type of breathing required for physical rest
lets us relax, focus, and concentrate in order to really rest.

So sit up straight in your seat. Uncross your legs. If your chair has
arms, lay your arms along their length. If the chair lacks arms, let
your arms fall loosely, the tips of your thumbs against the sides of your
chair.

Five Breaths Open the Gateway to Physical Rest

First breath: With your mouth closed, take a deep breath in. Feel your
belly rise outward as you breathe inward.

Now breathe out, slowly. Gradually but firmly push the air out of
your lungs, feeling it flow across your nostrils and back out into the
atmosphere. Don't be distressed if you sense a belch or some glitch
in your breathing. It's probably just some air moving around in your
digestive tract, readjusting to changed pressures. It's easy to feel these
glitches because of the tensions of daily life. They quickly decrease
with directed breathing.

Second breath: Again, take a deep breath in, this time feeling the
way the air moves inward across your nostrils. Sense your belly expanding, getting full.

What you feel is your abdominal muscles helping your diaphragm
expand the movement of your lungs. If you do not sense much
change, take a second to put your hand over your belly. Now let
the tips of your fingers feel the muscles push up with your inward
breath.

Next, breathe out. Do it slowly. Feel the air pouring through your
nostrils, first quickly, then gradually with diminishing flow. Keep
that airflow going, as you push more air out of your lungs. Don't be
surprised if you notice one of the nostrils has a lot more air going
through it than the other.

That's normal. The nasociliary cycle opens and closes nostrils at
different rates throughout the day and night, just one of thousands of

natural body cycles. If one nostril is completely closed, don't worry. What's closed now will open up soon unless you've got a cold or allergies.

Third breath: This time, open your lips. If you can, pucker them a little bit like you're ready to kiss. Pursing your lips lets you easily sense the air flow across your lips and tongue.

Breathing through an open mouth can allow you to move more air, just like you do when you're running or biking. Some people like to make a fair bit of noise when they breathe through pursed lips though others prefer silence. If you are in public and someone is watching you, you might want to breathe quietly.

Breathe in through your open pursed lips, feeling and hearing the air going by. As you breathe in, feel your breastbone move forward and a little upward. Next you should feel your rib cage moving *upward and outward* as you fill your lungs with air.

You want to make that breath deep because you want to really fill your lungs. Much of the lower region of your lungs is collapsed. It's the effect of gravity. Gravity presses down on the lungs and collapses their lower parts.

Breathing a bit deeper fills up more of the lungs. It also helps with what physiologists call V/Q, or the relationship between ventilation (air in) and perfusion (blood flow).

With each beat, the right side of your heart pumps blood into your lungs. The entire left ventricle of your heart is reserved to pump blood everywhere else, but the whole job of the right ventricle of your heart is to send blood to the lungs.

As the blood reaches the lungs, it eventually has to travel down from arteries to arterioles to capillaries, like water flowing from a river into a stream and then into a tiny trickle. Eventually the blood cells reach the spherical sacs called alveoli, which are specially designed to fill up with air. The oxygen in that air then gets picked up through the superthin walled capillaries of your lungs, filling the red blood cells inside your capillaries with the oxygen that keeps you alive.

If parts of your lungs are collapsed, the blood is not oxygenated.

Much of the blood going to the base of your lungs is then wasted.

However, that will not be so true if you learn to breathe deeply because that way you open up some of those collapsed air spaces. Fill the alveoli with air, and the blood coursing through will pick up the oxygen it needs.

That blood will then go in to the left side of your heart and get pumped everywhere else in your body, keeping you healthy. As you can see, it's really worthwhile to take a deep breath.

Now, as you exhale, let the air out slowly. Feel it flow down and through your mouth; hear your breathing quiet as the air moves out into the atmosphere.

Fourth breath: This time, count as you breathe. As you breathe in through pursed lips, slowly count in your head from one to four. If you think you're doing it too quickly, count as athletes do: one one thousand, two one thousand, three one thousand.

As you're breathing in, you will feel your abdomen expand outward. But now, pay more attention to the forward and upward motion of your breastbone and ribs. As your rib cage moves up and out, you will feel your back straighten, improving your sitting posture as you breathe. Your shoulders lift, rolling a little toward your back.

Exhale. Count down as you do, slowly going from one to eight. As you breathe out, feel the air going across your lips, and listen to the slowly diminishing sound. You may also feel a little tingling on the tips of your lips as the air flows by. Sense your shoulders descend as you breathe out.

Does this process sound familiar? It should. You did a version of this form of physical rest on day 6 of your sleep makeover when you went into that hot bath, letting your body core temperature open the sleep gate and send you toward a deeper, more restful sleep.

Fifth breath: Do as you did with your fourth breath. Breathe in to the count of four, out to the count of eight. Feel and hear the air coming in and out, opening up your lungs, filling your blood with the oxygen you need to live, move, and think.

This time, visualize the air as it moves. See it as it flows from your mouth, down your throat, into your windpipe. There it quickly flies down a hidey-hole of branching channels, each smaller than the last, much like the branching of a tree. These channels of air get tinier and tinier, until they extend all the way down into your spherical sacs of alveoli, where the blood pumped from the right side of your heart is moving, rushing by, trying to catch some of the oxygen that provides the energy that propels your cells to act, rebuild, and restore.

Fortunately, this process is remarkably efficient. Part of the reason it works so well is the amazing hemoglobin molecule. Hemoglobin acts much like an enzyme, grabbing the oxygen molecules from the air and holding them tight—until it needs to give them up.

It is all accomplished through a phenomenon called the Bohr effect, named after Christian Bohr, the father of the great physicist Niels Bohr. Niels Bohr helped develop atomic theory and quantum mechanics. During World War II, as a half-Jewish citizen, he was trapped in Nazi-conquered Denmark. At great risk he was flown out of Denmark in 1943 locked in the bottom of a Mosquito bomber. The Allies went to enormous lengths to bring Bohr to Los Alamos to aid the British and Americans in creating the atomic bomb. The air supply was so weak through the hours of his desperate flight that Bohr nearly died. Bohr was probably saved through the Bohr effect.

The Bohr effect makes picking up oxygen into a very quick process that operates through acid-base balance. Your body tissues are working away, using up oxygen, which increases their carbon dioxide levels and makes them more acidic. The more they work, the more acidic they become. The Bohr effect makes it progressively easier to offload oxygen from your blood into your more acidic cells. Similar enzymatic molecules, like chlorophyll, create the energy for most of our food supply as well as for our heating and transport. Oil, gas, and coal are the residues of dead plants, whose energy originally came from the sun. Thanks to our specially evolved enzymatic molecules, all that is required for nearly all the energy we use in daily life is sunlight added to water and air. Our survival comes from the extraordinary biophysics of molecules like chlorophyll and hemoglobin.

What You've Accomplished in Less than a Minute

Five breaths may not seem like a lot, but it is. With those five breaths you've

1. opened up the base of your lungs to better circulation
2. relaxed your body
3. concentrated your focus on what your body does in ways that let you both appreciate it and enjoy it
4. found a way to rest that literally takes seconds and can be used virtually anywhere and anytime

But you don't have to stop there. It's fun, and generally a good idea, to consciously breathe for more than five breaths.

As you keep breathing, concentrate on feeling the air go through your mouth, pharynx, windpipe, and lungs. Even better, try to see it. See if you can visualize that air flowing throughout your lungs, seeking places where it can pass to your blood cells the oxygen you need for life.

Don't forget to feel the movements of your abdomen, chest, and shoulders as you breathe. With time you'll do more than feel them. With time you'll actually see them too. You'll visualize your rib cage going up and out, sense the freedom as your chest moves forward, feel the actions that so easily and quickly bring atmospheric air down into the lungs.

As you breathe, you can also begin to try and breathe more slowly and evenly. You'll count the time of your breaths as you breathe in and out, making sure your exhalations are twice as long as your inhalations. That slow, measured push of air out from your lungs not only improves the balance between ventilation and perfusion but will increasingly relax you, calming you as your breathing becomes more efficient. It's no wonder athletes learn breathing as a critical part of their competitive training. With time, they are able to do more and more with less and less oxygen, often feeling great release as they move and play. Eventually such play becomes a form of flow, a way of attaining pleasurable experiences and peak performances.

Breathing, like most forms of physical rest, improves as you practice. It takes a little time. But it's a technique you'll be able to do for the rest of your life, so it's worthwhile getting good at it now.

DAY 9: PHYSICAL REST TECHNIQUE 2— MOUNTAIN POSE

My grandmother often said position was the thing in life. She was referring to social prestige, or what more currently is called socioeconomic status. Romantic couples usually see a different meaning. What few realize is that proper positioning of the body itself creates a state of physical rest.

Mountain pose is one position from the school of yoga devised by B. K. S. Iyengar, though many others have used it and its variations throughout the centuries. It's really simple. Mountain pose is just one way you stand up and breathe.

Practitioners of yoga may tell you that doing mountain pose properly will make you feel confident and strong, like a mountain. Our goals will be a bit less ambitious—learning to stand in a way that's restful, relaxing, and refreshing.

There are also other benefits to standing correctly. Mountain pose can improve posture. That's not a small issue in a country where youngsters and oldsters sit hunched in front of keyboards, monitors, TV screens, reading pads, and cell phones. It's not just our diet and forms of exercise that will make for shorter people, but the way we sit and stand.

About 80 percent of Americans complain of back pain. The health costs alone are in the many billions of dollars, the treatments often ineffective. There are many causes of back pain, but some problems can be prevented by learning to stand properly.

There's another advantage too. You'll also look good.

How to Do Mountain Pose

Mountain pose is all about alignment. You want to feel and look physically straight.

So please, stand up. Stand with your feet apart, pointing forward, at the width of your shoulders. Look straight ahead. Now, try to align your ankles, your knees, your hips, and your shoulders in an imaginary straight line. Imagine that line going from the floor through your body and up to the ceiling or, if you're outside, the sky.

Except for militarily trained people, many of us do not stand up very straight. We slouch a little at our midsection, tip forward our chests, curl our necks, and jut out our chins rather than tucking them in. So concentrate on seeing that imaginary line working through you, up straight from the floor and through your major joints. Once you do, within a few seconds you may start to feel straighter. Often you will also start to notice the solidity of the floor against your feet.

Next, notice the strength of your leg muscles pulling upward. You should sense your calves stretching, your knees and your thighs pulling forward. This sense of straightening will continue, moving upward through your hips directly on to your shoulders. Often this feeling of becoming straighter is accompanied by a sense of increased strength.

Roll your shoulders back, but just a little, as if you're starting to stand in a military pose. Tuck in the chin, but don't make it rigid. In mountain pose, you use your muscles and your breathing to relax your whole body.

Now, breathe in. Breathe as you've already learned, feeling your belly move forward, but focus primarily on your chest. Feel the air filling your lungs, causing your chest to expand upward and forward. As you breathe in, you'll feel your shoulders and scapulae roll backward, moving your chest a little more forward.

Next, breathe out. Do it slowly. You'll feel your shoulders come forward a little bit. You should hear and feel the air firmly moving through your nostrils back out into the atmosphere. Your ankles, knees, hips, and shoulders are still aligned in that perfect straight line.

All you need to focus on are two things—keeping your alignment straight, and breathing.

With each breath, you should feel the strength in your calves and thighs and the increasing expansion of your chest. Many people tell me they feel as if their upper body is getting slightly bigger, as they discover they are able to take larger, deeper breaths.

As you breathe out, make the exhalations long. Imagine you are cleaning up your lungs, both taking in and breathing out more air. The deeper, longer breaths should make you feel more relaxed. With time, they will also make you feel more alert.

Mountain pose is like many other techniques in this book, skills that with practice you can accomplish very quickly, often in one minute or less. When you don't have much time in your life, you want to use it well. With luck, as you relax using mountain pose you'll feel an expansion of time, as if what you've done has lasted far longer than it has. For many, doing a few breaths in mountain pose requires only about twenty seconds.

When to Do Mountain Pose

It's a good idea to try mountain pose as soon as you wake in the morning. Get up, and stand next to your bed. Taking a few deep breaths in mountain pose is a good way to start the day. It's even better to do it outside, under sunlight, as part of your morning brain-warming technique.

You can do mountain pose any time you choose. If you spend much of your day sitting in front of a computer or monitor, standing up for thirty seconds every hour or two can prevent some of the body kinks so many office and intellectual workers feel. Staying in one spot all day is exhausting, and it's not healthy. Blood pools in our legs and feet, increasing the possibility of superficial or deep venous clots. Joints stiffen if not used. The heart performs better if challenged by even the briefest moments of exercise, like standing up, since standing up takes about 25 percent more energy than sitting. Remember, you need to break up your day into different segments to give your body a chance to rest and restore.

Mountain pose is also useful in the evenings when you've been sitting for a while dining, reading, or watching TV. If you find it relaxing, you can also make mountain pose part of your sleep ritual.

Where to Do Mountain Pose

Any time or place where you normally stand is an opportunity to use mountain pose. It's a natural rest technique when you're standing in lines, such as at a bank or movie theater. Try mountain pose when you're waiting in line for airport security instead of worrying that the line is moving too slowly.

Personally, I use mountain pose whenever I enter an elevator. It's also a way for many to prepare for the workday. After you've dragged a heavy handbag across your shoulders when going to or from a lunch break, or when carrying several grocery bags from the car to the kitchen, mountain pose can act to reset your back, joints, and mind.

If You're with Someone Else

Sometimes people notice you're doing mountain pose as you stand next to them, though often they don't. At first, it's not easy to do mountain pose while engaged in conversation. With time, however, resting in mountain pose can be accomplished readily, especially if you are with someone you know well. It's fun to teach it to them as well. Mountain pose can make you feel better, and look better too.

DAY 10: PHYSICAL REST TECHNIQUE 3— GRAVITY POSE

It's amazing how relaxed gravity pose can you make you feel. After all, it's based on speed.

Your body is presently hurtling into the center of the Earth at a speed of 9.81 meters per second squared. That's fast. Even if you entirely neglect the effects of acceleration, it means you are falling at 22 miles per hour straight into the ground.

It doesn't feel like you're falling, of course. The mass of the Earth is pushing up with equal force. Your muscles resist your falling flat onto the ground, and the specialized muscle indentations around your veins, called valves, prevent the blood from pooling in your feet, stopping your descent both into unconsciousness and onto the floor.

Gravity pose is one way to actively and pleasantly resist the pull of gravity. Most physical rest techniques should automatically help the rebuilding of your muscle fibrils, tendons, ligaments, joint fluids, and venous valves that allow you to stand upright and stride across the Earth. Gravity is a powerful force that can be used to help you physically rest.

How to Do Gravity Pose

Gravity pose is even simpler than mountain pose. The first, most important, and most difficult part of gravity pose is to simply lie flat on the floor.

Once you're on the floor, join your hands by lacing your fingers behind your head. Next, cross your legs at the ankle, and raise your feet up from the ground. With your ankles still crossed, point your feet toward the ceiling.

Sense the power of gravity pulling you into the Earth. Feel it on the skin of your back. Sense it on your hips, your scapulae, and your shoulders.

Now breathe. Breathe in and out, exhaling slowly. As you breathe out, you'll feel your body sinking comfortably and gently into the Earth.

With each exhalation you'll also feel your legs sinking back to the ground. That's exactly what you want. As your legs descend with gravity you will feel your leg muscles relax.

You will also sense the muscles in the back of your neck relax, as you breathe slowly and comfortably.

Focus on how your back, legs, and neck feel as you do gravity pose. With each breath, sometimes with every moment, you'll feel yourself comfortably sinking, your muscles relaxing as they float back downward toward the earth. When starting gravity pose, breathe in and out at least three times. Get as much of your back's surface in contact with the floor as you can. As you breathe out, more and more of your back will touch the floor, relaxing you further. To get better at gravity pose, start it again when your feet have nearly reached the floor, lifting them back up to point toward the ceiling.

Physics can be fun. Gravity can help you rest. As you'll see, it can also help you sleep.

Where to Do Gravity Pose

The tough part of gravity pose is lying down. Finding a place to lie down can prove the most difficult part.

It's not hard to do gravity pose in your bedroom or living room. Having a carpet or rug beneath you is all you need.

Gravity pose can of course, be done on the mattress of your bed. As you breathe, you feel yourself comfortably sinking into the mattress, relaxing all the points of your back in contact with it. Many people use gravity pose as part of their sleep ritual, letting them pleasantly sink into sleep.

Doing gravity pose at work, however, may pose a challenge. I have my own office, which is carpeted. Yet many work sites have hard surfaces and not a lot of space in which to move around.

Even if your workplace has a smooth, soft floor surface, it's often worthwhile to buy a yoga mat that you can carry with you. Finding a free three-by-six-foot space will usually suffice. Getting your supervisor to allow you to use that space may be more of an obstacle.

If you do have trouble taking a minute or two away from your desk to do gravity pose, you might try telling your boss that gravity pose should help (1) relax you so you can work more efficiently; (2) prevent the back pain, neck pain, shoulder pain, and eyestrain that come from sitting in front of a monitor all day long; (3) decrease work-related repetitive task injuries, potentially leading to less time off work and lower health care costs for your organization; and (4) improve your mood and morale.

Of course, your boss may not believe you. Despite its advantages, gravity pose at work may prove a hard sell. People have yet to understand the benefits, indeed, the necessity of active forms of rest.

One way to convince your boss of the advantages of gravity pose is to do it very quickly—in a minute or less. Another, somewhat riskier, proposition is to let your boss try it out. It depends on your relationship with your boss and what you know about her or him.

There's plenty of data to show that exercise during work hours markedly increases work efficiency. People who exercise a half hour around lunchtime are more productive at their jobs, perhaps in part because they overcome the biological clock torpor of midday (see chapter 8, "Rest at Work"). Brief, active rests at work can improve work performance, social cohesion, and mood, even in very little time.

When to Do Gravity Pose

Gravity pose is useful when you're stressed out, brain fogged, or feeling physically tense, but it can be used anytime you want to rest, relax, and restore your sense of calm. It's certainly fun to do gravity pose in the middle of the afternoon or as part of your sleep ritual prior to going to sleep. Which leads us to the next physical rest techniques.

DAY 11: PHYSICAL REST TECHNIQUES 4, 5, AND 6—A BRIEF NAP, THE UNNAP NAP, AND A HOT, DAYTIME BATH

Naps restore you. Even a six-minute nap can improve memory and problem solving. Naps can act as the ideal pick-me-up in the middle of the afternoon or while at work.

Winston Churchill always tried to take a daytime nap. Many world leaders take "power naps." Before the industrial revolution, most of the population routinely took naps.

Why? Human beings are built that way. As you learned in the chapter on sleep, before artificial lighting from gas or electricity, people normally slept at night *and* during the day. Our body clocks decree it.

We can use body core temperature as a proxy for what our body clocks want us to do. When our core temperatures are going down, we sleep. When they're going up, we're more alert.

And when those internal temperature curves turn flat, as they normally do in the afternoons when we're not suffering jet lag or working the late shift, we rather readily fall asleep. Usually the best time of day for naps is between 1:00 p.m. and 4:00 p.m., with larks taking naps in the early part of that period, owls toward the latter part.

Our inner clocks tell us a lot of things, including when we can expect good or even just reasonably effective physical and work performances. One of the times we tend not to perform so well is in the early to midafternoon, when our temperature curves are flattening.

However, we can break through that sense of torpor, fatigue, and work inefficiency by taking a short nap. We don't even have to sleep. Just lying down and getting a rest can help a lot—and can increase mental and physical performance for the rest of the day.

However, research done in 2009 by Sara Mednick of the University of California–San Diego, a major nap researcher, argues that some amount of REM sleep is necessary if you want your naps to improve decision making and creativity. That usually means longer naps, according to Mednick, sometimes as long as two hours.

Naps are not for everybody. Some people are too wired to nap. Some people sleep so soundly they never want to nap. Some people who nap longer, generally more than twenty or thirty minutes, find those naps interfere with nighttime sleep.

And naps are not always a sign of health. People who must nap, especially in the morning or for several hours during the daytime, may be suffering from sleep disorders or other illnesses, including things as simple as a cold. Feeling sleepy is usually a signal from your body that you need to rest. If you don't rest, you don't rebuild easily and effectively. Rebuilding and renewing are always necessary, but the need is particularly acute when you're sick.

Viewing naps as silly, lazy, and wasteful is shortsighted. Quick, programmed naps can make people feel rested and mentally sharp, improve learning and memory, and if performed in the early afternoon, increase work productivity.

A Brief Nap

If people are so sleep deprived they must nap, they can nap almost anywhere, even standing up. Night-shift train conductors, as studied by Torbjorn Akerstedt of Sweden's Karolinska Institute, often take involuntary naps standing up—with their eyes open. Yes, it's possible to be asleep with your eyes staring straight ahead.

But if your goals are rest and refreshment, you will probably want to nap lying down. Here are a few basic technical requirements.

Equipment: A carpeted floor space of at least three by six feet and a pillow. A yoga mat or futon can be used on a hard surface or can be placed directly on a carpeted one. Wadded-up jackets or other clothing can stand in for pillows. A couch or sofa is often more comfortable for a nap. If you are at home, you can nap on your bed. Whatever spot you find comfortable is a good place to nap. I never thought I could easily nap on the floor until I tried it. As there is a couch in my office, I usually use that, though I find it only marginally more comfortable than my carpeted floor.

For optimal efficiency, a night mask is also helpful. If one is not available, a wadded-up cotton terry washcloth, or almost any kind of cloth, can be used to cover the eyes against unwanted light.

Nap technique: Once you've found a safe place where you feel it is comfortable to nap, lie down. Stretch your legs and arms. Put your arms along your sides or comfortably across your chest. Place the night mask over your eyes.

Now, focus on breathing. Breathe in to the count of four, out to the count of eight. Visualize the air moving easily in and out of your lungs.

You can even give yourself a mental rest (discussed in the next chapter) by visualizing a scene that will help you sleep. Sure, go ahead and count sheep, if you like. Yet it's easier to imagine yourself slowly and comfortably sinking into a deep piled carpet, descending further and further with each breath.

If simple breathing does not sufficiently relax you, do gravity pose. As your back sinks into the earth and your legs fall with the force of gravity, feel your muscles relax. Sense the relaxation as you breathe in and out.

If you want to nap and still don't fall asleep, that's okay. You nap so you can rest. As you learned in the previous chapter, people in light sleep are often completely unaware that they have achieved sleep.

Rest is the goal. If you don't quickly achieve effective sleep, do not fret. There are also ways to achieve the UnNap nap—a nap that can make you feel quite rested without falling asleep, as you'll read about below.

How Long Should I Nap?

How long to nap is quite controversial. Some researchers, like Sara Mednick, consider a two-hour nap productive. They say that people go through a long sleep cycle during such naps and come out in light sleep, making it easier to wake up.

The problem is that such studies are generally done on healthy undergraduates or graduate students. Things are different in the clinical and work worlds. You often can't predict what stage of sleep someone will wake up in. The circadian rhythmicity of sleep is different during the day than at night, and long naps can easily land you in stages of sleep from which waking up leaves you with the ghastly feeling of sleep inertia, the terrible slowness and lethargy that come when waking from deeper phases of sleep, even relatively light stage 2 sleep. Sleep inertia is a disaster for workers who need quick reaction times, like those laboring in hospital emergency rooms or nuclear power plants. It is a problem for the military, where naps are used when preparing for the many night operations performed by soldiers in the Middle East and South Asia. Sleep inertia is a particular headache for commando teams, which sometimes get little sleep over sixty to eighty hours, and may help explain why specially supplied caffeine gum has become so popular among soldiers in Iraq. Moreover, many working people find that if they take long afternoon naps they can't sleep that night or that their nighttime sleep is shorter and less restful after they nap.

You won't run into most of these problems with short naps. To avoid sleep inertia and obtain a useful daytime nap, set the amount of time for your first nap at fifteen minutes. To not go over that period of time, use a timer.

It can be a kitchen timer. It can be your cell phone—if your calls are blocked. It can be the alarm on your watch. Just set it to ring at fifteen minutes, and put it close enough to you that it will wake you up.

Many people prefer naps longer than fifteen minutes, but if you normally don't nap, start with these shorter naps so you can learn to nap effectively. Using an accurate timekeeping device allows your biological rhythms to aid you in your quest for rest. People who start napping using a timer often find they can more easily control how long they nap.

Just as approximately a third of the population wakes up within five minutes of their morning alarm, people trained by timing devices can often internally set the lengths of their naps. If you get used to a fifteen-minute timed nap, you may quickly train yourself to nap for fifteen minutes without *any* timer. Body clocks can be so efficient that people can often allow themselves ten or twenty minutes for a nap and easily wake up without a watch or timer. This precision may be the result of the extraordinary influence of our twenty-four-hour clocks, which are constantly monitoring daytime physiology, and their interaction with some of the shorter internal thirty-, sixty-, and ninety-minute rhythms that exist in so many individual cells.

Politicians know this trick. As described by Simon Sebag Montefiore in his *Stalin: The Court of the Red Tsar*, one of Stalin's major henchmen told his family he could nap for fourteen minutes. They timed him, and that's exactly when he woke up.

Short naps can do more than increase work productivity. They can also make people feel good.

When to Nap

Nap times depend on body clocks. Most folks prefer to take their naps between 1:00 p.m. and 4:00 p.m. Generally, larks prefer to nap in the early afternoon, owls in the late afternoon.

If your job or work practices allow it, try naps at different times and see which time works best for you. Some find they can nap anytime during the afternoon. However, to resync your body clocks, it usually pays to take your brief naps at the same time each day.

Where to Nap

If you have a lot of free cash, there are companies that for a high fee will take you into a specially constructed van and let you nap in the middle of the workweek. (They've consulted me about their methods and market, so I know this is true.) Many airports have special new nap facilities, while corporate downtowns are now studded with special spaces offering rental naps. Silicon Valley corporations, with their sometimes unusual work schedule requirements, have become very progressive and have set up nappatoria, where their constantly operating workers can take a quick snooze whenever they like. Some law firms are making it easier for round-the-clock workers to nap in specially equipped offices and conference rooms.

Yet you have to know what your company allows. A report in 2008 in *The Wall Street Journal* noted that a survey by Circadian Technologies, a consulting firm with close ties to many Fortune 500 corporations, found that of the companies reporting, 52 percent had suspended or reprimanded workers for napping, up from 38 percent in 2002.

That might explain one of the newer iPhone apps—the iNap@ Work. While you snooze at your desk, the iNap converts your cell phone into a work noise generator. Throat clearing, typing, computer, and paper shuffling sounds are automatically produced by your iPhone, convincing perhaps some supervisors and coworkers that you are deeply engaged in labor.

Napping at home is a different story. Lots of sleep-deprived Americans take their naps on the weekends.

Often they don't have a choice. Their bodies want or require eight or more hours of sleep, and they get only six or seven hours during their weeknights. If your body needs the rest, it will often just go and take it. Weekend afternoon naps of up to an hour can often do a lot to make up for chronic sleep deprivation during the workweek.

The UnNap Nap: What to Do When You're Not Allowed to Nap

Many employers will reprimand or fire you if they catch you napping on the job. In many cases they have excellent reasons. You don't want to be in a train whose conductor is nodding off at the controls.

But if you have a job where napping could be helpful to your work performance yet is still frowned upon, let me propose a solution—the UnNap nap.

Impossible, you think? The trick is to use the techniques developed a century ago by Edmund Jacobson, a remarkable psychologist and medical doctor who was one of the first to show that the autonomic nervous system can be placed under conscious control. (An excellent description of Jacobson's technique is found in A *Headache in the Pelvis* by David Wise and Rodney Anderson.)

Jacobson called his technique paradoxical relaxation, and he often liked to demonstrate its effects. He was said to make himself so calm that within a couple of minutes people began to think he might be dead. When giving interviews he was so relaxed that others would quickly become nervous.

The paradox of paradoxical relaxation is that you get relaxed without trying to become relaxed. All you do is pay attention to muscle tension.

Though it takes practice, here is a quick way to try paradoxical relaxation for your UnNap nap

Close your eyes. If you're in a work situation where this will be unfavorably noticed, put your hand over your eyes as you close them, and leave your fingers over your eyes.

Now, feel the motion of your eyeballs. They are not stationary, even when your eyes are closed. Our eyes normally are moving a lot while we are awake, sweeping our environment to notice what is happening around us.

Do you feel your eyes jiggling even though your eyelids are closed? That's entirely normal. Your eyes also move when you are asleep, and the normal beginning of sleep is generally noted in the sleep laboratory by observing small rolling movements of the eyes.

Now, sense the different levels of muscle tension and motion in your left eye. (Use your right eye if left-handed.)

Do you feel a little more tension in the left lateral corner? Or is there more muscle tension in the middle of your eye? Just pay attention to the different levels of muscle tension, as if you're mapping it across the eye. (If your eyes jiggle too much, concentrate on the muscle tension differences of an eyebrow or a patch of your forehead.)

Now, find the spot in your closed eye with the greatest muscle tension. Compare its level to that of another part of the eye right next to it.

Feel the muscle tension in that tiny part of your eye. Feel its intensity. Don't try to modify it. Don't increase it, decrease it, or try to change it. Just notice it.

A further paradox of paradoxical relaxation lies in that paying attention to one specific muscle usually relaxes muscles elsewhere across the body.

This won't happen suddenly, at least for most people. You want to try paradoxical relaxation two to three times a day for at least a minute each time, just focusing on the different levels of muscle attention in your eye.

After a while, if your attention begins to wander, try to visualize what those tiny, tense muscles look like. Imagine you can feel and see the tensed-up muscle fibrils. Muscles work by small overlapping filaments of actin and myosin proteins sliding in and out of each other something like sets of teeth in a comb. Try to visualize these small muscles of your eye. Feel the tightness, sense the muscle fibrils sliding deeply into each other or knotting up.

After a minute or so, open your eyes or remove your hand from your eyes. If someone asks you if you're okay, explain you were just trying to think very intensely for a little while.

Once you practice the UnNap nap, you will generally find you get more and more relaxed the more you try it. Try your first UnNap naps in the evening or late morning, when most people are relatively sharp. As you get better at them, you can try UnNap naps in the mid- to late afternoon, a time you may really want to nap.

You can then focus on muscles other than those around your eyes. Once you start to track your levels of muscle tension, you will start to feel and notice different muscle groups all over your body.

Got a tense shoulder? Maybe you twisted it last night carrying the old newspapers out to the recycling stop. Focus on that shoulder during your UnNap nap. Or if you find the eyes difficult to concentrate on, try sensing patches of an eyebrow or part of your forehead. Looking at a monitor all day usually makes one or more parts of our eyebrows tenser than others. I often find that I can start an UnNap nap by keying in to the muscle tension in my lips or across the width of a finger.

Sometimes it's fun when doing your UnNap nap to map all the different levels of muscle tension in your body, seeing the most tense muscles as deep purple, the lesser levels of tension in orange, the least tense in green. As you notice the different levels of muscle tension, your whole body should feel more and more relaxed.

UnNap naps can make you feel more relaxed within a minute, but you usually can get far more relaxed within three to five minutes. And if you do want to sleep, very pleasant, brief naps can begin with UnNap naps. Even if short, they can be startlingly refreshing.

Why Nap?

Napping, as you can tell, is controversial. It should not be. Napping is a normal part of human design. Even with the problems involved in napping at work, a poll of American adults carried out in 2009 showed that 34 percent nap in the afternoon.

Here are some reasons to nap:

- Napping might save your life. A six-year study of Greek adults completed in 2007 found that those who took thirty-minute naps at least three times a week cut their heart attack risk 37 percent. Even fewer heart attacks occurred in napping middle-aged men.

- You'll work better. One NASA study found naps averaging twenty-six minutes improved work performance on some tasks 38 percent.

- You'll feel better. Your mood often rises when you take naps.

- You'll deal with people more effectively. Napping does more than improve mood. Lots of patients tell me that they can deal with others far more happily if they have had a nap. It's not just that their fuse is longer, it's that they also feel more sociable. This result rarely shows up in short-term productivity studies, but it can have a big impact on the effectiveness of an organization over time.

- People are chronically sleep deprived. Napping improves motor skills, which means fewer traffic and work accidents.

- Adolescents need naps. With the average adolescent sleeping six to seven hours at night and needing nine or more hours to really learn and function, even short afternoon naps will improve homework performance and grades.

My bottom line—most people can benefit from fifteen- to thirty-minute afternoon naps. Often the increased work productivity is well worth it. But the improved mood and sociability can prove very helpful and restful in the rest of life.

A Hot, Daytime Bath

Lots of people feel too wired to take a nap in the daytime. If you don't want to nap—or even take an UnNap nap—you can take a hot bath instead. The Victorians often did this, and it went on to become a significant perquisite of upper-class life.

We already discussed a hot bath as a technique for falling asleep. However, a hot bath can also be used as a simple, effective form of daytime physical rest, and does not require you to sweat. Try this hot bath technique.

Start the bath, aiming for water that is hot but not scalding. As you run the water, stand in mountain pose for thirty seconds to a minute as you check the temperature.

When the water is hot and the bath half full, plunge in. Rest your buttocks comfortably on the tub's bottom. Breathe slowly and evenly. Focus on the heat spreading up and through your body. Once you feel fully relaxed, get out and towel-dry.

At first a daytime bath may make you feel very relaxed and a little sleepy. Yet if the bath is short, you may soon perk up. To wake up more quickly, breathe in the bathroom using mountain pose.

Baths in the daytime do take time. You have to undress and dress, run the water, and dry off. However, a four- to eight-minute daytime bath can also break up and reconfigure the pattern of your day.

You want to feel the rhythm of the day, the internal excitement that comes with easily moving from activity to rest. A hot bath can act as a fine reset, especially when you've been upset, anxious, or exhausted that day. But other forms of physical rest may be more accessible and can rest you anytime, anywhere you choose.

Summary

Physical rest is a form of active rest in which you use simple techniques that focus on your natural body processes. Physical rest is quick, easy, efficient, and effective and can be done almost anywhere and anytime.

Here are the quick techniques of active physical rest you've now learned:

deep breathing

mountain pose

gravity pose

brief naps

UnNap naps

hot daytime baths

All of these techniques can be done easily, but they improve with practice until they become skills you can use anywhere you live or work. As you learn to physically rest, you can feel:

more relaxed

more alert

calm

centered

able to work

more in touch with your body and its internal rhythms

You will also be able to teach others how to physically rest. It's easy, and it will help people you care about learn to rest whenever they like.

chapter 4

MENTAL REST

Mental rest is a different form of active rest than physical rest. In physical rest, you concentrate on the body and its processes. In mental rest, you concentrate attention on something beyond your body.

The focus can be on visual images, sounds, or mental imagery. With focus comes power. Using the mind to concentrate on a single item can powerfully affect the sympathetic and parasympathetic nervous systems, changing blood pressure, heart rate, and body temperature. The effects on mood, anxiety, and creativity can be even more profound. To understand a little more about mental rest, please answer these questions:

1. Mental rest involves
a. focusing my mind.
b. stopping thought.
c. becoming very physically inactive.
d. physically tiring myself in order to get some rest.

2. While I mentally rest, my brain
a. slows down.
b. goes to a different systemic and metabolic configuration.
c. becomes less able to concentrate.
d. makes me feel sleepy.

3. The best place to mentally rest is

a. the kitchen.

b. a movie theater.

c. in front of the television set.

d. any place I choose.

4. A good time to mentally rest is

a. when I need to.

b. when I'm frustrated.

c. when I'm tired.

d. when I'm all wired up.

e. all of the above.

5. The ability to mentally rest changes with

a. time of day.

b. overall physical health.

c. how I shift from rest to activity.

d. what I did just before I rest.

e. all of the above.

6. To feel in a state of flow I should

a. not notice the passing of time.

b. feel concentrated and focused.

c. act with a certain degree of skill.

d. feel a sense of challenge.

e. all of the above.

Answers: 1. a; 2. b; 3. d; 4. e; 5. e; 6. e.

People are exhausted these days, though they often are not sure why. There are plenty of reasons. They're sleep deprived, for one. They're trying to juggle family, work, and social obligations and still enjoy activities that bring them simple pleasure. Frequently, the activities they think will make them happier, like watching television, don't, and they feel more exhausted. They often try to do too many things at once. A 2009 study of texting while driving showed about half of sixteen- to twenty-four-year-old Americans and one-fifth of all American adults text on a wireless device while operating a vehicle. We have now taken our fascination with multitasking to seriously dangerous levels; truck drivers' accident rates go up 2,300 percent while texting.

Unfortunately, a lot of people have lost an understanding of the brain's need to focus. Mental rest is all about focus—concentrating the mind. Most human achievement requires close and sustained focus. What can be achieved by such focus can be phenomenal.

Tibetan monks can sit in loincloths in subzero weather and voluntarily move their body temperatures up and down. People can undergo major surgery while under self-hypnosis, a form of active rest in which physical relaxation is high, mental concentration is high, and attention to the overall environment is diminished. People who actively meditate claim that they experience mental and physical energy bursts after ten or fifteen minutes of meditation.

But you don't need to be a champion focuser to enjoy and benefit from mental rest. Mental rest happens when you focus on something beyond your body processes. Ultimately it can let you see the world in a heightened, completely different way.

Mental rest is about mental re-creation—reconfiguring your mind to quickly and easily obtain a sense of relaxed control. You're concentrating, but you are also relaxed. The big stuff doesn't bother you because you're engrossed and engaged in what you're doing now. Your actions feel seamless, unitary, often playful. With practice, mental rest allows you to control your consciousness and your life more effectively, even for seconds at a time.

The first technique of mental rest you'll learn is self-hypnosis. It works by having you pay attention to your body in ways that relax you *and* focus your concentration. Once you have gained this inner feeling of relaxed concentration, it can deepen your sense of what you can and will accomplish.

Lots of people don't want to try self-hypnosis because they think they will lose control. They've watched too many old movies. With self-hypnosis, you *gain* control—of your physiology, your attention, and what subjects your brain wishes to consider. Self-hypnosis activates many different areas of the cortex and lower brain structures in a quiet, rhythmic way. The result is the relaxation response, where the body decreases its output of oxygen and does more with less. The parts of the brain activated by self-hypnosis are quite different from those activated in paradoxical relaxation, the technique you learned in the previous chapter with your UnNap nap.

When doing almost any task, like watching a car speeding down the street, different parts of the brain are activated, while others are turned off. The default mode network, also called the task negative network, tends to turn on when we are not goal directed but instead are meditating or just daydreaming.

In contrast, self-hypnosis is a process involving a lot of mental concentration and focus. You can feel very, very relaxed doing self-hypnosis, but your brain is doing a lot.

If you have any misgivings about learning self-hypnosis, please take this quiz:

1. If I learn self-hypnosis, then
   a. all kinds of people can hypnotize me.
   b. I can turn into a bear whenever I choose.
   c. I can gain a state of relaxed concentration.
   d. I'll never fear airport security lines.

2. Self-hypnosis
   a. requires that it first be taught by a hypnotist.
   b. will make me excessively salivate every night.
   c. is inexpensive.
   d. is something I can learn by myself.

3. People use self-hypnosis to
   a. physically rest.
   b. help them sleep.

c. help them achieve many different goals.

d. all of the above.

Answers: 1. c; 2. d; 3. d.

Generally self-hypnosis is easy to use, quick, and endlessly adaptable. Let's learn how to do it.

DAY 12: MENTAL REST TECHNIQUE 1— SELF-HYPNOSIS

In a world that bombards us with information, the art of selective attention becomes increasingly important. That means that self-hypnosis is a skill you'll want to add to your toolkit.

Self-hypnosis is just one form of self-controlled concentration, but it is a powerful one because it allows you to focus in a very relaxed way. When there's already too much going on all around you, it pays to concentrate.

Just as there are hundreds of ways to relax, there are dozens of ways to self-hypnotize. Here is one simple technique you can use.

The Eye Roll

Over the years much research has been done on the eye roll. A few people have great eye rolls, many others don't. Some have the kind that make self-hypnosis faster and easier, while others have particular difficulties. There are even those who get goofy doing the eye roll.

Genetics plays its part. If your parents had good eye rolls, you'll probably have a good eye roll. Good eye rolls make it easier to do self-hypnosis quickly and thoroughly.

The biology of the eye roll has not been worked out, but the psychology has been studied extensively. People with a good eye roll, according to its most famous practitioner, the late David Spiegel (his son continues his research), tend to emphasize feeling over thinking.

They are more suggestible than others and more prone to have powerful imaginations.

Learning self-hypnosis is actually pretty easy. The eye roll is nothing more than looking straight ahead and then rolling your eyes up to the top of your head like you're staring at the ceiling.

Try it now. Sit straight in a chair. Point your head straight ahead, looking at an object directly in the front of your visual field. Next, while keeping your head pointed forward, look straight up.

Yes, up. Hold the pose for a little bit. It if feels a little strange, you're probably doing it right. Most of us don't point our heads straight ahead and then try to look at the ceiling unless we want to self-hypnotize.

Next comes the tricky part. With your eyes looking straight up, slowly, *slowly*, close your eyelids. While you're looking straight up, your eyelids are coming straight down, like the curtain in the theater.

Believe it or not, that's the hardest thing you will have to do to learn self-hypnosis. A really good eye roll will show lots of white on your eye as you close the eyelids. An even better eye roll will form into an A pattern, where the eyes move inward and stay that way as you close your eyelids.

None of these phenomena are things you can see, of course. I, for example, possess a perfectly awful eye roll, according to my hypnosis instructor. I still enjoy doing self-hypnosis a lot, though it took quite a bit of practice to learn.

Try the eye roll two or three times, sitting down in a comfortable chair. If it's uncomfortable, just try it for ten seconds. Feel your eyes behind those closed lids. You'll sense them jiggling, with quite a lot of muscle tension, as they look straight up.

Now, do your eye roll again. Your eyes are fully closed, but you still *keep your eyes looking straight up.*

No, you won't see anything except the light that normally filters in through a closed eye. If you wanted to, you could tell if it was brightly lit or dark where you're sitting, but that's not your present concern.

Instead, you want to concentrate on keeping your eyes looking up while your eyes are closed.

Now, take a deep breath, just as you learned while deep breathing (physical rest technique 1). Again, breathe in to the count of four, out to the count of eight. Feel the air pass across your lips. Hear the sound of the flow of air, its rising and ebbing.

Your eyes are still closed. You're still looking straight up.

As you breathe out, you'll feel the back of your neck start to relax. Your neck's relaxation is a natural part of exhalation, but it also makes you begin to feel a sense of calm release.

With your next exhalation, feel that relaxation spread from the top of your head down to the base of your neck. Breathe slowly. The deeper and longer the exhalation, the greater the sense of relaxation.

Sitting, breathing, your eyes closed, you can begin to feel that sense of relaxation spread farther into the rest of your body.

In your next breath, feel that sense of relaxation pass from your neck down into your shoulders. As you breathe deeper and deeper, your sense of relaxation will eventually start to feel like a wave of warmth as the muscles progressively relax, region by region by region.

With your next breath, feel that relaxation move down through your shoulders and into your chest.

Every new breath can deepen your sense of relaxation as you feel the muscle relaxation moving slowly down through your body. Soon you may feel your belly relax. Next, the wave of relaxation moves down to your thighs and legs. Eventually with each succeeding breath the sense of relaxation slowly spreads from the top of your head, down your neck, into your shoulders, down your chest, sliding into your abdomen, then flowing down your legs all the way to your toes.

If you don't feel relaxation going throughout your entire body, don't fret at all. For many, achieving that level of deep relaxation takes a bit of time. What matters is whether you feel even a tiny bit more relaxed doing self-hypnosis than you did before. Remember, your brain is quite active and concentrated as you do self-hypnosis, in contrast with your body, which gets more and more relaxed.

To gauge whether you really are more relaxed, you will want to come out of your controlled, self-hypnotic state. Do three things: (1) keep your eyes up at the top of your head; (2) take a deep inhalation; (3) open your eyes.

When you open your eyes, you should feel more relaxed. Most people feel a bit more present and calmer. Some feel a calm they've never experienced.

> ### If the eye roll makes you uncomfortable:
> Some people have trouble doing an eye roll. It makes them feel uneasy, even a bit weird. It's a small minority, but if that's the case with you, relax. Chances are you will be able to induce self-hypnosis through deep breathing and need not trouble with the eye roll.

FAQs About Self-Hypnosis

Where should I practice self-hypnosis? Anyplace you feel comfortable and safe and have a pleasing seat or position to use.

How often should I practice self-hypnosis? To get pretty good, you want to practice for one to three minutes three times a day. As normal alertness peaks in the late morning and early evening, either or both times are good periods to practice self-hypnosis. Another time to try is right before sleep, as many people use self-hypnosis as part of their sleep ritual. Self-hypnosis is an effective form of self-induced body relaxation.

Once you've practiced self-hypnosis for a few days, you should be ready to put it to some enjoyable uses.

Self-Hypnosis for Simple Relaxation

Self-hypnosis is a form of highly focused concentration. It makes sense that often the easiest thing to start focusing on is concentration itself.

First, you start by focusing on something simple and calming—a pleasing word. People generally prefer to choose their own words to focus on. If you're not sure what word to choose, try focusing on *home* or *peace.*

As you breathe in, hear the word in your head. Listen to your word in your own voice. Don't focus on anything except that word. If you focus exclusively on your single word, you're already learning a new form of concentration. You will later be able to put that power of concentration to many good uses elsewhere.

Concentrate on your word for two to three minutes. With a little practice, you should feel more relaxed and calm.

Now you are ready for the next step. To really feel refreshed, it usually pays to do some visualization. Visualization means seeing something in your mind. What you will see is up to you.

Normally we visualize lots of things throughout the day. Much of brain activity, particularly during rest, probably involves simulations of what can or might happen to us. However, the type of visualizations I'm talking about here are directed, explicit, and entirely under your control. You will probably want to visualize many varied things, some of which are extremely pleasant. When using visualization as part of self-hypnosis for mental rest, you want to concentrate on images that are calming, refreshing, and relaxing. You don't want to focus on things that provoke deep emotions, just feelings of calm, pleasure, and simple happiness. Such feelings will tend to relax your body more.

So now you've accomplished your eye roll, and you are breathing easily and deeply. You have felt your neck and chest relax, your muscles lose their tension, your mind become very aware of your breathing and its sense of inner control.

Now try these two sample visualizations.

Visualization 1

You're dressed in a swimsuit, comfortably walking along a beach. You've never visited this beach before. Its shape is a near-perfect crescent. The sand is remarkably fine, like the grains at the end of a barrier island, cool to the touch and kissing soft. The beach's color is a shade of sand you've perhaps seen once in your life, a light, pastel pink.

Looking out from the sand, you see that the water is strikingly clear. Peering down at the sand lining the bottom of the water, you see it is formed into miniature peaks and valleys, like a perfectly realized monochrome landscape. The waves roll on quietly, making a pleasant tinkling sound on reaching shore. The surf is slow and even, flowing horizontally in waves whose tops form clear rolling lines—energized, circular forms that appear like the contours of a painting by Kandinsky.

Do you feel a little more relaxed? Please keep on walking. The sand is soft and cool, tickling your little toes as they touch the ground. You stare for a while at the play of light as sunlight skips off the waves. Beneath, the water's floor is a suddenly bright chalky pink, the sand formed into pure color.

It's remarkably pleasant walking the shoreline, following the curl of the beach as it gently reaches out to rolling surf, but the water is simply too inviting. Quietly you ease your way into the surf. Effervescent coolness strokes your calves as you wade into softly bracing water. Looking at the sand below, you see that your legs are rolling through a transparent liquid that seems to brim with bright white light.

If you're feeling a little adventurous, duck your head in and dive underwater. As you open your eyes, wavy lines of light flash across the sand, as if the sun were personally searching the bottom. You watch the flashes for a while, pure squiggles of light.

Then, off to your left, you see something brightly colored. Small fish, quicksilver and fluid, dart into your field of vision. Before you can catch them on your retina, they escape.

Quickly you turn your head. Now they are in front of you. You see them rolling along the gently flowing current, watching you, staring at the new visitor. Now, slow and friendly, they approach. You watch their fins lazily flick to the side, the small indentations of their mouths seated beneath wide, surprised, gold-flecked eyes. You notice the strange but remarkable body colors, all rich and full, red, yellow, blue,

green. Light ripples across their sides and backs, their fins momentarily sparkling as they lazily swim before you. You can finish this first
visualization by nodding to the fish, and then slowly let yourself float
away toward the shore.

If this beach is not for you, try this second script.

Visualization 2

You have not been able to concentrate as well as you would like. A
dozen things roll around your mind, worrying you through much
of the day. At this point the idea of a vacation seems beyond your
imagination. A few minutes alone, without work, kids, partner, or
parents, feels like a real vacation to you right now. When you look
out from your work space, your best view is a long office corridor
lined with a featureless gray carpet that some previous corporate
decision maker thought looked professional. *Institutional*, that's the
word you'd use. Heck, they could use this office space as the set for
a horror movie.

You walk out into the corridor. It's empty but inviting. No one else
is present.

You decide you're going to do something you may not have done
for many years: you are going to flip high in the air.

It's been a while, so you take your time. First, you put your hands
down on the ground, tucking your legs behind you. Quickly, you feel
your legs kicking up and out. Their springing power amazes you.
In less than a second you're flying up through the corridor. You're
moving so quickly you wonder how you'll ever land straight. But land
you do, upright, and with barely a wobble.

You've done it. You feel strong, supple, ready to go.

It was so much fun, you do your flying somersault again. And
again. Soon you are springing down the whole length of the corridor.
It's so easy you can't bother to stop. You dare yourself to kick higher,
fly higher, stay longer in the air. It's sad when you see the end of the
corridor rising up. You slow down, finish, and look back.

Your body feels strong. Your mind feels refreshed. Mentally, you feel like a gymnast as you open your eyes.

Visualization 3

Here's a chance to make your own visualization. Close your eyes, and travel to an uninhabited, natural place you'd really like to visit. When you've arrived, start the most enchanting hike you have ever known. It can be hiking up through mountain forests, curving along a glassy black beach, or roaming through fresh-smelling, ripening grain under a giant sky. Start your journey, and see where you can go.

Self-Hypnosis for Sleep

Doctors spend long stretches of their lives getting little sleep. So when they do get time to sleep, they really want it to count.

Over the years I've discovered that many of my colleagues, even ones so sleepy they don't think they can reach a bed, use self-hypnosis to fall asleep. Some accomplish it simply by breathing in and out, concentrating on words like *peace* or *sleep*. Others prefer different forms of self-directed imagery.

During rapid eye movement, or REM sleep, different parts of the brain turn on and off. The different systems that are activated and inactivated are quite unlike those of waking consciousness. For one thing, much of the parietal lobe gets turned off. The parietal lobe includes parts of the brain that help tell you your position—where your body lies in three-dimensional space.

Turning off your sense of position can create rather curious results. Here's an example.

You're in a dream, talking to a friend in your home. The next moment you're still talking, but suddenly you're in Montreal, speaking with someone else.

You have no idea how you moved from your house to Montreal. You've no clue how or what provided transport or even why. Yet it does not seem to matter to you in the least. You move around the world effortlessly with the ease of a novelist writing your own story, and everything about it feels natural.

Another effect of turning off position sense during sleep is that nearly 99 percent of us have dreams in which we fly. These common flying dreams can now be used as entry points, allowing us to visualize images that quickly let us enter sleep.

One of my presleep maneuvers—used by my friends too—is to begin such visualizing as soon as my back or side rolls onto the mattress. As I wait for sleep, I concentrate and see myself floating.

Sometimes I'm a few inches above the mattress as if some antigravity device is keeping me up. I feel the air dreamily curving up and around my body, cooling me. Somehow that flowing air is the sweetest cushion I've ever known. If I'm not quickly asleep, I try to float a little higher, about two feet below the ceiling. I can't remember what happens after that because I just fall asleep. Later in the night, when I do have dreams, they sometimes involve flying.

Many of my colleagues who like to influence dream content often visualize flying before going to sleep. Sometimes they move as if they're positioned on a glider. Others try flying with their arms and legs outstretched, a bit like Superman but without fears of kryptonite or Lex Luther. They see themselves passing over the land, curving up and down, sideways and back, speeding up or gliding down to take a look at what lies below.

Some visualize traveling through the Alps or Rockies, waving their hands at vertical crags of ice and stark blue-white glaciers. Others feel a rush of wind and suddenly feel themselves lifted, like a hang glider catching a sharp updraft. A few cavort around Paris like characters in the movie *The Red Balloon*, getting a bird's-eye view of anything they want to see and feel.

Usually flying visualizations are nice ways to fall asleep. Visualizing flying may also pleasantly change your dream content.

DAY 13: MENTAL REST TECHNIQUE 2— FOCUSING THE EYE

Physical rest focuses your mind on your body. Mental rest lets you restore and mentally re-create by focusing on elements beyond your body. With self-hypnosis, you have now learned a form of relaxed concentration that can be applied to almost any mental image.

Many of your images will be self-generated—you do what you want to do and go where you want to go. Yet mental rest, with its enhanced sense of calm, vitality, and alertness, can also be quickly obtained by focusing on the environment around you.

The problem is, there's a lot out there, usually too much to focus on. Paying attention to the whole world can prove overwhelming, even if appreciating the vastness of what is around us is often exhilarating. You'll see how when you do spiritual rest technique 4, contemplating suchness.

But first it's easier to pay attention to something specific, small, and doable. For many of us, it's good to start by picking out one single aspect of the natural world.

Focusing on Nature

Most of us can see or look at trees or plants. There are several palm trees waving in the distance outside my office window, so I normally focus on a frond of one of them. Up north, the branches of oaks and maples are easy objects on which to concentrate.

First, just look at the frond or branch—nothing else. Focus your eye on it, keeping your attention on it for twenty seconds. At first it may not feel like much time. However, if you are really focusing, seconds can feel like quite a long time.

You've probably looked at that frond or branch in the past and immediately looked away. You saw nothing of real visual interest. But look again, slowly and carefully. Start by classifying what you see. What color is it? Is it uniform throughout? Does it have elements of yellow-green or brownish green? If there are leaves, how long are they? How would you describe their shape? What is their size? How many of them are there?

Next, If there is any wind, observe the movement of your branch or frond. Does each leaf flow in the breeze in the same direction, or do they move separately? Are the leaves' movements coordinated, the waves rippling in a breeze? Do they move jerkily and rigidly or smoothly through the air?

After you've done this classifying, try to see the frond or branch as a whole. Don't try to describe it in words or mathematically measured lengths. Just try to see the frond as a totality. Can you catch all of it, altogether, in your mind? Or do you have to separate the frond into words, or separate images, that only later you put together into a single, whole picture?

Take a look at your watch or cell phone. How long have you been looking at that frond? Have you been able to keep other things out of your head?

What Your Brain Is Doing

When you're first observing your frond, you are classifying, studying, analyzing. You are using words, mathematics, comparisons with other images in your memory.

For the brain this is an amazingly active process. You are activating the visual cortex, your senses of movement and measurement, many of your memory stores of the plants you've seen and known. Though the brain is very active, you yourself can feel far more rested.

That's because you're thinking of just one thing. You're focusing. You're not thinking of the outrageous cost of the latest phone bill, your long commute time to work, or why your kid won't practice the instrument you slaved to buy her. Nothing is getting in the way. If it does, you quickly return to looking at the frond. Now you're paying attention, sharpening your concentration, focusing on a very small part of the natural world.

Just noticing can allow you to see a lot more over time. The better you get at seeing the world, the more you will see. Concentrating on your perceptions is like other forms of learning. Different connections within the brain are strengthened. Your increased perceptual capacity then becomes available to you throughout your life.

Your fuller sense of focus pulls you away from all those random thoughts and lets you take greater control of your mind. Then, as you try to simply appreciate all that is before you, further changes take place.

As you look at the frond, nothing else, seeing it without trying to describe and define it through language, your brain shifts into a different mode. You are no longer classifying but contemplating. You can feel a sense of internal quiet, of peace and rest.

This does not always happen as you move from seeing to contemplating the frond. For many, the act of seeing quickly becomes interesting and exciting. They begin to notice all the movements of the plant, sense its different forms in space. After a while they can even feel what those actions are like.

Sometimes people tell me that instead of looking at the frond, they try to imagine themselves as the frond. They sense the breezes, the temperature, the changing reflections of light, even the taste of the air. Within ten or twenty or thirty seconds they've moved away from themselves and considered the world from a thoroughly separate, restful vantage point. Some begin to feel part of a much larger universe, engaging that sense we call the spiritual. And they're doing it quickly and simply, in the spirit of play. You want to make rest a truly active type of fun.

Where to Focus the Eye

You can focus on any living thing or any image you like. However, it's easy to start with some natural being or object that you actually have a chance to look at.

If you're stuck inside a windowless office, a photograph of something that you find beautiful can be used. However, it's nice to focus on some object that *is* natural. Any living thing can be visually interesting, though it's best to start with something that remains visually stable. Your dog may be fascinating, but plants stay in one place better. An orchid on your desk, or a small potted plant like a cactus or jade plant, can make focusing the eye accessible wherever there is a source of light.

When to Focus the Eye

Practice aids most skills. It's certainly true of the many kinds of mental rest. If you can, it's good to practice focusing the eye at least two to three times a day. All the time you need for these attempts is twenty to thirty seconds each. At first, even such a brief period may seem a long time. Later, as you do more focusing the eye, you may find that the time seems to hardly last at all.

When you reach that point, pat yourself on the back. It means focusing is becoming a flow experience, a skill where you use your mind to create challenges that give you control over your experience and sensation of the world. Feeling that sense of flow can change the dullest day into something interesting and entertaining.

If you have very little time for focusing the eye, try taking a quick peek at nature before and after meals. Don't be surprised if focusing the eye starts spilling into your other senses. The more acute you become in your seeing, the more acute you may become at hearing or taste. You can do more than taste a fine meal. You can smell it, see it, and feel the changes of textures of the foods until the whole experience becomes strong and rich. That's what focusing can do.

DAY 14: MENTAL REST TECHNIQUE 3—WALKING TO MUSIC

IPods are popular for a reason. We love to hear music. We love to move to music.

Give kids a chance, and they'll move to virtually any song—and they will move as a group. Humans move to music in groups. Witness

Woodstock, any large rock concert, or a rave. According to Oliver Sacks in his book *Musicophilia*, other species are different. They move to music, but they move alone, not together as a group. We are rhythmic in our bodies and souls, and our rhythmic actions resonate with others.

Walking is also a rhythmic activity, one we love to do. To many, walking is an art.

Now you can start to make walking musical.

How to Walk to Music

You can use an iPod, a cell phone, or any electronic device that plays music. However, it's just as much fun, and allows more flexibility, to simply use your own head.

Most of us hear music in our heads, often a good part of the time. Perhaps we use that music to tune our minds. Not uncommonly, we can track our moods by what's playing in our heads.

To begin walking to music, choose two tunes or melodies that you really like, one fast and one slow. Hear each in your head for at least twenty seconds.

Now find a spot where you'd like to walk. It might be a garden, a street filled with shimmering storefronts, or a park in the middle of a city. However, if you're like many of us, it will probably be something less lovely, like a parking lot or an office corridor that could double as a set for a remake of *The Shining*. To walk to music, you need some space but surprisingly not that much.

Concentrate on the fast tune. Next, hear several different instruments or voices playing and singing it. Now, walk to its beat.

At first, walk in a straight line. If you see work colleagues or supervisors, nod appropriately, but concentrate on that melody. Move with it.

Walk to the fast tune for twenty seconds. Next, turn around.

This time, walk to the slow tune. It may feel weird for your calf and thigh muscles, but try it out, moving in time with the song in your mind. Most of us have a natural rhythm of walking, which you will recognize when you walk with someone else. The music playing in your head, especially the slow melody, may not fit your natural walking rhythm.

Move to that song anyway. Hear the music and feel it in your chest, your thighs, your calves, your ankles. Get your shoulders to move in rhythm to the beat.

As you walk, you'll begin to see that going from point A to point B creates one of the great opportunities for what is perhaps our most natural common action. The human body is a walking machine. We have walked the deserts, the mountains, the plains, the high mesas. We have even walked to the North Pole, to pretty much every land destination on this earth. We have and will continue to walk to our food, to our homes, to the people we love.

Yet when you walk to music, you do something more than just walk. You concentrate on rhythm, meter, song. In this way the act of walking becomes like dancing.

You've walked for twenty seconds to fast music, then twenty seconds to a slow tune. Notice how different your muscles feel with each.

Now, listen to the different tunes. What are their moods? Happy? Tragic? Silly? Gleeful?

As you walk, you can feel those emotions yourself. Just as you can with an iPod, in your mind you can choose the tune, the rhythm, and the mood.

The next time you walk to music, choose both a melody that is happy and a melody that is sad. Walk in time to each. Notice how you feel moving with each song.

Do your muscles feel the same? Do your shoulders swing more when the song is happy? Do you sense a narrowness in your movements when the melody is sad?

If you do, you'll recognize that much of your consciousness is at least partly under your own control. When you walk to music, in a few brief moments you can feel the music, beat, rhythm, and emotion and the changes they bring to your body—all in quick, startling simplicity. Now you can use walking as more than a utilitarian travel device; it can begin to feel like an adventure—a restful adventure. By tapping in to natural rhythms that reflect the innate communication modes of your body, music gives the mind a sense of restoration, of increased capacity to do more of what you want or need to do. Just moving the body can itself rest the mind.

When to Walk to Music

It's often a good idea to try walking to music at the beginning of the day as well as near its end. Music and motion can then become bookends to your waking day.

Sometimes you won't have that luxury. So when the day is tense and you find your work hard and stressful, take a minute or two to walk down a corridor. If for purposes of propriety you must have a destination, go to a restroom or walk to music while moving up and down a stairwell.

When you need a quick boost of energy, walk musically to a fast, happy tune for a minute or, if you have the time, two or three minutes. Walking fast musically is the first of several Power-Ups (described in chapter 8, "Rest at Work") that can get you going very quickly. Many of these techniques activate areas of the brain that also evoke pleasure, increasing dopamine inside what are called the reward circuits of the brain.

Lunchtime is also a good chance to walk to music. If walking with someone else, be sure the other person can handle your brief lack of attention, or explain that you need a short time to think about something.

Where to Walk to Music

Most mental rest techniques can rest you nearly anywhere. You can walk to music in the kitchen, in the living room, or rhythmically walk the apartment building parking garage on your way home.

Walking to music does not take long to do. You can enjoy significant benefits from just a minute or so—though you can do it for far
longer if you wish.

Walking to music makes music feel more a part of you. Do it when
and where you can. You'll have the pleasure of feeling music in your
head and throughout your whole body.

DAY 15: MENTAL REST TECHNIQUES 4 AND 5— EAR POPPING AND GARDEN WALKS

When I was a child a movie based on a hit play appeared called *Stop
the World: I Want to Get Off.* The movie was a bust. Yet today I hear
that phrase more and more.

It's hard to get off this world. Just ask any astronaut. A few wealthy
individuals have spent time at the International Space Station after
paying many millions and undergoing tests and physical training that
might scare professional athletes. One of them, entrepreneur Richard
Garriott, told *Time* magazine in 2009 that growing up as the son of
an astronaut, he thought becoming an astronaut was something he
could easily do. He was shocked when told as a teenager that his poor
eyesight would "forever" prevent him from becoming an American
astronaut. Garriott was determined he would get into space. He paid
a reported $30 million to reach the International Space Station in a
Russian spacecraft.

My advice, however, is that no matter how hard things get, you don't
need to jump off the planet. What you need is a reset button, preferably
one you can use wherever you want. If you're in the middle of work or
family turmoil, you might need to stop and take stock. And if you're
surrounded by hostile relatives, crazed neighbors, or a CEO on the
warpath, you may have only a few seconds to push that reset button.

In stressful times, you need a quick reset button that's restful too.
Ear popping fits the bill. By stopping and suddenly restarting your
brain's perceptual apparatus, ear popping can rapidly provide you a
way to both quickly slow things down and look at them anew.

Ear Popping

Sometimes you have to drown out all the noise. With ear popping, you simply put both your index fingers in your ears deep enough to stop outside noise. Leave your fingers there for ten seconds if you have the time, five if you don't. If you're in a place where it's socially acceptable, also close your eyes.

First, listen to the silence. Some people, when they try ear popping for the first time, hear a deep rumbling. Others may hear their own heartbeat. Both experiences are normal. However, what you really want to focus on is the decreased sound, the decreased noise. If you can, focus on hearing *nothing*.

After five or ten seconds (it may feel longer), noisily pop your ears by rapidly pulling out your fingers.

Now, open your eyes. Look around.

Concentrate first on immediate visual perceptions. Where are you? What colors do you see? Are they deep, fixed, saturated, or are they foamy and light filled? Notice each color in turn—green, blue, white, whatever is most prominent in your environment. Notice the light in the room. Is it coming from windows or flowing from artificial lighting? What is its color, its strength, its brightness?

If you are indoors, look around. Look at forms. What are their shapes? Are the objects in your field of vision square, circular, rectangular? Are they chairs and tables, desks and lamps?

Next, focus only on sounds. What do you hear? How many different sounds are there? Can you figure out where the sounds are coming from? How many are generated from something nearby, how many by objects far away?

Finally, notice the people in the room. Try to feel their presence rather than focus on their visual appearance. What are they there for? What are their motivations? Do you have some inkling of their agendas?

It will probably take you only a few seconds or a minute at most to notice all these things about your environment. But in this brief time you have reset your perceptions so that you can more fully pay attention to what is around you.

You can ask many different questions about your surroundings, yet the answers may appear nearly instantaneously. The brain works quickly, and once the right questions are in place it answers them with surprising rapidity.

With ear popping, you take a brief bite out of time. For a few seconds you look at where you are and what you're doing, taking note of what is around you and what is happening to you.

With practice, you can use your ear-popping reset very quickly and efficiently. Yes, it will look odd in some social situations. Tell anyone who asks what you were doing that you were trying to clear your ears It's true—ear popping will cause you to pull natural wax from your ear canals. Unless you really want to, you don't have to tell anyone that you're simultaneously clearing your mind.

There are many ways to reset your immediate perception. Zen masters may slap acolytes on the cheek, forcing them to pay attention to what is happening, now, in the eternal present. You can reset yourself much more pleasantly.

After ear popping, the world should look a little brighter and clearer. The brain appreciates that you pay attention to it and how it sees the world. Once your perceptions are reset, it's easier to engage in many other techniques that can rest and restore you.

Garden Walks

In some religious texts the first image of paradise is of a garden. As the Chinese say, nature is high, wide, vast, and deep. To feel a sense of natural wonder, you do not need access to a formal garden.

To find a garden that will rest your mind, you only need to find something from the natural world. Once you find your piece of the natural world, it only takes a few brief moments of observation to provide you with mental rest.

Lots of folks tell me how on the job, even during travel, they miss contact with the natural world. Sometimes you need to bring some of nature with you. However, it can become a special experience if you go to visit it.

Finding a Garden

Sometimes, to find a garden all you need to do is look up.

Across the street from the condominium where I live stand the requisites of modern urban life—electric lines and telephone poles. They are so ubiquitous and immobile many people do not notice them. When asked to draw a landscape or the appearance of their street, most people leave them out.

I don't spend much time looking at telephone poles. I tend to observe them only when a natural interloper perches on them.

The burnished steel electric lights across the street are often the physical rest points of ospreys (fish hawks). Perched high on top, ospreys hold their brown and white wings spread out evenly from their body's center as they peer down searching for prey, emitting high sparks of sound to signal their partners.

Stranger to the eye are the small round balls of tillandsia bromeliads that periodically latch onto the telephone pole wires running across the nearby avenue. The small gray and green filaments look like some organic version of electric life, a displaced bit player from a sci-fi movie made odder by the strangeness of their placement. Seeing a complexly tangled, gray-green sphere happily thriving on a naked high wire is bizarre.

There's no earth, no dirt, no evident form of sustenance. How does that thing *live* there?

It survives because it's an epiphyte. Epiphytes attach themselves to other structures, living on the water and nutrients that rain down on them from the air. Not just vagabonds, flimflam artists, and financial con men live on air.

Nature is almost everywhere you look. Unless you're working in a semiconductor fabrication plant, it is really hard to keep living things out.

Instead, you can bring them in.

Walking to a Garden

When you think of a garden as any place where there is something natural to look at, it becomes easy to find a garden. Lichens often appear on the tops of thoroughly dead rocks, and tufts of grass sometimes peek right out of the hardened brickwork surfaces of office buildings. Try as we might, we cannot keep nature down.

Walking can be an inherently restful action. As you will see, it can also act as a flow activity involving challenge, skill, and the pleasurable suspension of time.

A mental rest can be quickly obtained just by walking outside to look at nature. Social psychological experiments performed at England's Essex University in 2007 showed that walking in nature is much more powerful at improving mood than walking through an urban mall. Perhaps for evolutionary reasons, we like looking at green things. Fortunately, even "barren" cityscapes are inhabited by dozens of different forms of plants.

Many appear uninteresting until you really start to focus. Lawns are everywhere in the United States, but how often do you look at the grass?

There are dozens of different kinds of grasses. All have different structures, colors, aggregated forms. There is the phosphorescent green of the New England countryside and the pale brown-green of the woolly Florida lawn, filled with varieties of grass considered weeds elsewhere in the country.

If you walk past a home, you may witness an unplanned garden of many different types of flora. This time, take a few seconds and look at one such spot carefully. Try to focus on a place you never looked at before. Notice its structure as well as its variety, the inherent rhythms of its patterns of growth.

If you can see those patterns, patterns that make life possible, you will usually notice something beautiful. Much of the art of living beings is formed at scales we don't often see, but almost everything alive is filled with marvelous, visually stunning structures. This is also one of the pleasures of art. The patterns created by an expert Chinese or Japanese calligrapher may transmit words that convey a banal message, but the forms of the characters themselves will remind their readers of the flight of geese, the flow of waves on the ocean, or birds dropping down through the air. Chinese art critics first started lauding such naturalistic calligraphy about eighteen hundred years ago. Even the abstract paintings of Rothko or Pollock, supposedly representing their personal unconscious, often powerfully remind people of natural patterns and colors.

If you can't go outside, look for nature inside your home or workplace. Go make a formal, unannounced visit to a coworker's jade plant or cactus. Touch the smooth, small leaves, feel the waxy cuticle, its reflective, water-resistant surface. If no one has a plant, bring one of your own.

Don't worry about killing your plant. Some cacti are really hard to murder. I'm an inveterate plant neglecter, but my small office cactus has still managed at least a dozen winters with only grudging, mainly irregular intervention.

If you have a nice corner office, look out. Even fifteen seconds watching a cloud can make you sense its dynamic motion, its volumetric, three-dimensional depth. If you watch them on a breezy day, clouds may seem alive. They should. The most recent belief is that clouds are caused (in part) by cosmic rays. Just watch what they do. Shade trees on a street can change their patterns of reflection with each passing cloud.

You can also carry with you natural objects no longer alive. In cultures that pay great attention to nature, pieces of nature are brought inside to remind people of the season. The Japanese and Finns sometimes bring small leaves, flowers, or branches to their workplaces and homes. The pieces remind them of the whole, of the vastness of nature that exists both inside and outside where they sit.

Walking is rhythmic and inherently relaxing, but it becomes more so when there's a clear destination. One brief adventure is to find some living or other natural object and to place it in a spot only you know. Artists often play a related game, putting a small painting or drawing in an odd or obscure public space. You can put your natural object on a lawn, in a stairwell, or on the bookshelf of someone's office. Go to it sometimes, and see if it is still where you put it. Sometimes it will move. If it does, see if you can put it somewhere else. You can then conduct a dialogue with someone who also enjoys paying attention to your joint environment.

Walking in a Garden

The small pleasures of walking to a garden increase when there is a real garden to visit. Even in the deepest concrete urban canyons, parks can be found.

Walk to them if you can. Often a break from work or family can be obtained within five to ten minutes. As you walk, look around, noting all the different sorts of life around you.

Before you reach the garden, imagine how it will appear today. Will it have changed from when you last visited it?

It should. Nature, including your body, is constantly, cyclically changing. Seasons, temperature, light, and the depredations of animals such as humans modify everything we look at. Many a gardener adores watching a plant grow.

When you are walking in a garden, wherever you look you will find something new. Breathe deeply and slowly as you move, training your eyes, ears, and nose to notice what is around you. It's inherently restful to the mind to contemplate green living things. It's inherently animating to the mind to see the changes of nature, shifting through the days and weeks.

Walking through nature makes it possible for the everyday to not feel everyday. Several different British studies demonstrate that people who live near green spaces live longer. This is sometimes ascribed to the increased ability of people to walk and exercise and to their increased sense of social community and social rest, the subject of the next chapter. Yet much of the benefit may come from the mental rest that comes with being outside, in nature.

Summary

Life is rhythmic and inherently musical. So is rest.

Mental rest lets you use your mind and body to concentrate on the world around you. You can do mental rest quickly and easily and just about everywhere. Its main limits will be in your imagination, which will improve as you practice different techniques of mental rest.

In this chapter you've learned:

self-hypnosis

focusing the eye

walking to music

ear popping

garden walks

Now that you know what they are and how to do them, you can use these mental rest techniques whenever you like. You don't have to use them all, but they can be consistently tried out in various environments with surprisingly fast results.

Mental rest allows you to relax your body and mind, to connect with the world around you, to feel a part of the natural world. It can lower your blood pressure, train your eye, and lead you to feel bouts of joy in the strangest, most stressful of settings. Mental rest techniques can also be used to calm and reset yourself when there is nothing calm around you and to reacquire a sense of understanding and purpose in situations you'd otherwise wish to leave.

There are many different ways to rest. Put together, they are more powerful.

chapter 5

SOCIAL REST

We are social animals. We live and survive through our social connections. The power of social rest can fortify us, preserve us, amuse us, and give us purpose. Different forms of social rest frequently do a great deal to keep us healthy. Often they save our lives.

Our social interconnectedness is everywhere. There are nearly seven billion of us on the planet. In some places, like Chandni Chowk in Delhi or the streets of Kowloon, so many of us live together outsiders can't imagine how there's enough physical space to pack us all in.

And our species is getting more social. The world is becoming increasingly urban and increasingly connected. Distance, at least on our planetary scale, no longer affects communications that much. You can now watch the same Harvard lectures in Cambridge or Datong or immediately speak in real time to supersized visual monitor versions of your colleagues on any inhabited continent.

Why are human beings so social? We're built that way. Social connectedness is not just one of our many strengths, it is also part of the structure of our brains.

Still, if you are a little confused about what social rest might be, you can take this quiz:

1. Social rest is aided by
   a. social integration.
   b. social support.

c. giving and sharing.

d. all of the above.

2. The more social connections you have, the longer you'll live.

   True   False

3. The stress response seems to be biologically mediated by social support.

   True   False

4. Social connections have to be long and deep to be biologically useful.

   True   False

5. Social support can take virtually no time at all or last a lifetime.

   True   False

6. Resting socially precludes resting mentally.

   True   False

7. Whether I die of a heart attack is affected by how I spend time with friends.

   True   False

8. Whether I survive cancer is affected by my friends.

   True   False

9. Social support is as useful to survival as

a. avoiding all smoking.

b. never being obese.

c. never having high blood pressure.

d. all of the above.

Answers: 1. d; 2. true; 3. true; 4. false; 5. true; 6. false; 7. true; 8. true; 9. d.

I often talk to people who are near the end of their lives. They frequently discuss their vacations. A few talk about their money and social position. Many talk about their jobs. Yet what do most of them talk about and value most? Their relationships. The people who care for them, the people they cared for and care about.

Our social connections are so deep they sometimes get a bit spooky. Many times people will start to pick up the phone and find they're being called by the person *they* were about to call. Lots of people sense the feelings, state of mind, and personality of others. Such thoughts often occur right in their presence. Sometimes they occur when separated by great distances.

Many years ago a friend of mine spoke of a strange occurrence. Suddenly he was seized with worry for a cousin. He knew something bad was happening to this man right *now*.

He had not spoken to this cousin in years. He did not feel close to him at all. He just knew, deep in his gut, that something was really wrong.

My friend called his cousin's number. He was not there. A family member went to find him. There he was, working under the chassis of his car. The support blocks had collapsed. He was pinned down, crushed to the ground. His family came out and pushed until they freed him. The ambulance arrived quickly. His cousin survived.

My friend asked me what the scientific mechanism was that allowed him to know from thousands of miles away that his cousin was dying. I provided him the answer I increasingly give through years of medical practice: "I don't know."

What I do know is that these types of experience are frequent. People tell me they immediately need to call a close relative or friend, only to discover the person is ill, in the hospital, or dead. Others know who they will marry though they have never met the person. Loads of ordinary people tell me of such psychic phenomena. They often wonder out loud if they are crazy.

Based on my clinical experience, the vast majority are not crazy in the slightest. Most of us feel deep connections to others, often in ways we cannot express in language. Frequently such connections appear to defy present-day scientific explanations for communication, the passage of information, and the nature of physics.

We feel deeply connected to others in ways we cannot even name. In terms of our health, survival, and self-esteem, we are also profoundly connected. Social rest helps us live. The research on social rest and social connection really got going in the late seventies and is going strong today.

Berkman and Syme

In 1979 two Berkeley researchers, Lisa Berkman and S. Leonard Syme, published a follow-up study of 6,928 young to middle-aged adults. They were members of the Human Population Laboratory, a random population sample living in Alameda County, California. The two scientists decided to look at social integration and its effects on health.

They defined social integration as specific social ties—marriage, how many and what kinds of contacts people had with relatives and friends, and whether they belonged to community and church groups. No matter how or where they looked, social integration affected survival.

The increased survival margins were especially large for cardio-vascular disease, but most health categories were affected by the level of social ties. For people who were isolated, with fewer social engagements, the relative risks of death were 2.3 times greater for men and 2.8 times greater for women as compared to those more connected. Statistical controls were then added to survey most known disease risk factors.

After the Berkman-Syme paper appeared, social support research took off. Studies were done in country after country, with remarkably robust and similar results. As Sheldon Cohen at Carnegie Mellon wrote in one well-regarded review of the research that appeared in the 1990s, social support had a major impact on whether you died of a heart attack, whether you got depressed, whether you survived a cancer diagnosis or fought off infectious illness.

Other researchers went further. They asked what elements constitute effective social support. Is it just the arithmetic number, the lists of social connections, or does psychological and social meaning play a role?

Meaning certainly does play a role. It matters whether people perceive they are being supported and whether they provide support to others in turn.

The next question was why it all works. Why does social support aid survival? Some of the more recent research theories put forth are that social support

1. helps you cope
2. helps encourage more effective health behaviors (like not smoking)
3. provides stability
4. gives you a sense of control
5. improves self-esteem
6. gives you useful information and advice
7. plugs you in to a network that helps you function

These theories have all received some empirical corroboration. As an example, studies in 2009 show that diabetics with increasing social support also demonstrate increasing glucose control and that even "slight" social connections matter in positive ways. But that social support provides major psychological and social benefits does not tell you *why* it works. Others are trying to understand in biological terms why social support is so helpful.

Understanding the biological mechanisms of social support is very early in its development. Here are some of the things that researchers in biology have looked at since 2004.

The hypothalamic pituitary axis (HPA) is a huge neuroendocrine circuit that helps control the stress response, along with many adaptive responses. Social support activates the HPA, in part by cutting back dramatically on large cortisol increases. In studies done by Kenneth Kendler, reported in 2005, the rates of depression, especially in males, are decreased through social support. In other research, oxytocin, the hormone that increases one's sense of warmth and emotional connectedness, is itself increased through social support. Animals that have more support from their fellows have better immune responses and less nasty responses to infection.

However, our understanding of how the brain communicates and coherently acts with other organs through neurotransmitters, hormones, and direct nerve connections remains in its infancy. How most human biological systems communicate and interact remains largely unknown. We know that there are at least one hundred trillion different organisms, mainly bacteria, viruses, and fungi, living on or inside every person. Yet how these different organisms interact with our 10 trillion cells is at present unclear. One indicator of these different organisms' importance is that perhaps 8 to 12 percent of the human genome originated in retroviruses. Not until 1983, when the most famous retrovirus, HIV, was discovered, did anybody have a clue about what retroviruses did in people.

The body is an absolutely gigantic information machine. Knowing what causes what will probably take researchers a rather long time and will lead to a much better understanding of how its systems interact.

Still, what is known is that social support helps you psychologically, socially, and biologically to survive and enjoy each day. It also helps you in another way. Social connection helps you rest.

Social Touch

Social support has too many benefits not to use it frequently in daily life. The real question is how to do it.

One way is through the concept of social touch. The great novelist E. M. Forster once wrote, "Only connect." The Beatles sang, "I get by with a little help from my friends." Social touch lets you do that.

Many people desire the experience of connection. We crave physical and emotional comfort. We want to know somebody cares for us. Others want to feel that we care about them. Humans crave reassurance, itself a powerful mediator of medicine's placebo effects. We desire information and advice we can use. We want to feel embedded in a social network larger than ourselves and our immediate family. In the most basic social terms, we want to both give and receive.

Social touch lets us do that by "touching" others. We use our capacities for communication to let others know that we are there, and, more significantly, that we are there for them.

There are many ways to touch others. We can go over and talk to them. We can send snail-mail letters. We can send oral messages through friends. We can also use the Internet.

Social Networking and Social Support

Lots of people wonder whether Internet connections enhance or hinder social support. Certainly there are more social connections online—but are they all healthy?

Some clearly are not. Kids who text hundreds of times a day, including in bed, tend to get less sleep and appear more irritable, as my friend Gaby Bader noted in research done in 2008. Aric Sigman, a fellow of the British Institute of Psychology, noted in 2009 that face-to-face communications involve very different brain circuitry than virtual ones. He quoted Duke University research from 2005 that found that in the previous twenty years, with the rise of virtual communications, the number of people who said they had no one with whom to discuss serious personal matters increased from 7 to 25 percent. In 2005 Sigman pointed to other research correlating television use between the ages of twenty and sixty with increasing

Alzheimer's disease, further highlighting his fears about a more online society.

Yet when radio first became popular, many people wanted to ban it from cars because it would distract drivers. Humans have many different ways of communicating and interacting. The Internet has been a tremendous boon to people who are isolated, particularly through illness. Its use in education is only beginning. As for social connection, it remains to be seen whether quickly texting friends increases or decreases HPA stress responses. My belief is that the Internet is another layer of technology that can connect lots more people than many previous technologies. It will connect them superficially or deeply depending on how it is used. What's clear is that its uses will increase and that its health effects need to be carefully studied.

Gradations of Social Connection

Research shows that not all social connections are alike. Depth and intensity matter.

Most of us want to love and be loved. However, there are often difficulties with our deeper emotional connections.

People fall in and out of love. Children change the tenor and dynamics of a couple's relationship. Work comes to dominate and often interfere with the balance of our emotional life.

Many couples fail as couples because they try to be everything to each other. Though sometimes wonderful, such comprehensiveness is hardly necessary. Human social connections can be less close and still be very helpful.

Please take the time to write out a few lists. When you get to ten individual names on each line, you can stop.

1. people I can talk to anytime for personal emergencies
2. people I trust to speak about intimate secrets
3. close friends

4. people I like to talk to

5. people I wouldn't mind seeing more

6. people I can go to a movie with

7. people I can watch sports with

8. work colleagues I talk with easily

9. acquaintances I like

Look at the different lists. You should be able to see that social connections exist on many different levels.

There are people you can tell intimate secrets to but may not consider close friends or may not even like well. There are people you like and always enjoy visiting but hardly ever get to meet. There are folks you like to go out with, but the emotional and spiritual world you inhabit are as familiar to them as Mars.

Social connections exist across a wide gradation of activities, depths, and interests. You want to use and enjoy them all. You should also recognize that your purposes in using these connections may be very different.

You also know that the people you contact and see will change. Some people who are hardly acquaintances now used to be close friends. Someone you met briefly two months ago may a year from now become your lover. Your movie-watching buddy may one day turn into a business partner.

Social connections are multiple, fluid, and infinitely adaptable. That's one reason why social rest can be so much fun.

We are born talkers. Most of us love to converse. Many of the quick techniques of social rest involve talking and all the neuroendocrine, stress-hormone reductions, and psychological and social benefits that human conversations entail. Yet these connections count in more than social ways. Most researchers argue that social connections are at least as significant to your rate of survival as obesity or whether you smoke. The longest-lived population in the world, Asian American women in metro New York City, have wide and deep social connections. Talking to someone you care about and who cares about you can provide you a profound sense of rest and inner safety.

DAY 16: SOCIAL REST TECHNIQUE 1—MAKING A SPECIAL CONNECTION

Look at list 1, people you feel you can call in a personal emergency. For day 16 of this rest program, choose someone on that list and call him or her.

You can do it on a landline, a cell phone, on Skype or other videophone hookups. Talk to this person for at least a few minutes. After the usual chat of where and what and how both of you have been doing, tell the person that she or he is important to you. Ask if you can call, hopefully anytime or anyplace, if a crisis or emergency strikes.

Some will ask you why you are bothering to ask such a question. It's obvious you can call them whenever you like. This person is your spouse, parent, child, best friend, close friend—there's no need to ask.

Say thank you, and ask for the privilege of this special connection anyway. Your brief call lets the person know that she or he is important in your life. Social connection is about giving and receiving. Receiving a message that they are considered important and trusted makes many people feel better while some will feel honored.

And if this person refuses to take such a future call? Find out why. The reasons may surprise you. Though some might be "too busy," others, you may discover, have major problems of their own—medical, social, or financial—and do not feel they can fully help you in critical situations. With this added knowledge, you can see what you can do to help them.

Sustaining Special Connections

Making a special connection as a form of social rest means that you are not only making a contract for help in a crisis, but you are also building your ongoing connection to this person who might help you. If possible, you want this connection to that person to be both regular and frequent.

The reason is that such social connections provide sustenance and aid that make dealing with crises easier, quicker, and more comfortable. The stress-response data argue that social connections themselves prevent future problems. If you have people who you know will help you, chances are that whenever you deal with stress, you yourself will *experience less stress.*

Knowing there is someone out there who will help, at any time or any place, is a great relief to most of us. Meaningful social connections between people who care for each other provide a sense of security and inner peace that makes social rest highly useful.

What to Do

Make special-connection calls at least once a week, trying a different individual on your first list. As you speak to each person, make a record of contact phone lines and addresses that you can conveniently keep with you wherever you go.

Emergencies crop up when you don't expect them. That's part of what makes them emergencies.

Just as you never know when crises will appear, you don't always know who will be available to help. That's why you want to make your special connections list fairly broad.

And current. The people on your special connections list are people you will want to talk to well beyond periods of severe distress. They should be people you see as having the sense and judgment to help you when you really need help.

Since we are deeply social animals, we need help most of the time. Social rest is about that giving and receiving.

So try and contact the people on your list frequently, even if it is for the briefest of times. You may want to talk to them for two or three minutes every week or every month. You want them to know you care about them and what has happened to them. If you can help them, it makes it simpler, and more effective, for them to help you.

As people move in and out of your life, connections with relatives and close friends will change. Try not to let most such contacts lapse. As we grow older, we learn different things. One thing we tend to learn is how important relationships are to the sense of meaning in our lives. Such understanding makes people want to keep many relationships going while recognizing they will often change in character.

DAY 17: SOCIAL REST TECHNIQUE 2— VISITING A NEIGHBOR OR COWORKER YOU DON'T KNOW WELL

There are lots of people we think it would be fun or useful to know. Sometimes they are right next door.

Much of present-day communication is electronic. However, even the best and biggest visual monitors always leave something out. Some of what is lost is what psychiatrists call "relatedness," what many people call their "animal feel."

A huge amount of information and communication occurs beyond spoken language. We react to clothes, hair color, gestures, height, facial symmetry, smell, and a few thousand other nameable and unnameable traits. Often we find we like someone we have just met though we are hard-pressed to justify why or how.

Some of this information processing is unconscious (we have no idea where it comes from), and some of it is preconscious (we may eventually remember or figure it out). We may like a stranger because she or he reminds us of a former teacher we revered or a famous movie icon or has the same kind of sexy walk as a former boyfriend or girlfriend. Much of our like or dislike is both unconscious and physiologic. One example: our immune system has a lot to do with who does or doesn't smell nice. Studies started in the late nineties and continued through 2008 demonstrate that we are more attracted to those who smell *different* than we do. If we can believe evolutionary biologists, that's because our immunity-defining histocompatibility antigens are expressed through smell, and it's better to find mates outside our own restricted gene pool.

So pick out someone at work or in your neighborhood who gives you a nice "animal feel." Go over and introduce yourself, even if you've met before. Do it a little formally so the person will know you are doing a bit more than saying hi.

Then ask a few questions. Ask about something that should be of mutual interest, like what this person has heard about the new boss or, when meeting your neighbors, how they prefer to water the lawn or the best time to put out the trash. If such subjects do not lead any further, consider that many folks have a favorite conversational topic—themselves. If they don't want to talk about themselves, reveal to them something you like about yourself, and see what happens. If conversation falls flat, ask about climate, weather, or sports.

The social touch visit need last no more than three to five minutes. If it was in any way a positive experience, try to repeat it within the next week. If it was not positive at all, ask yourself why. The answers may tell you quite a lot about your workplace, your neighborhood, and perhaps yourself.

The Pleasures of Preindustrial Societies

The French are not always easy to love. One old French joke declares, "When the creator had finished making France, she was amazed by what she had done. The climate was salubrious, the air and water clear and sweet, the landscape beautiful, the land bountiful. Things were so wondrous she looked around and thought, perhaps she needed to balance all that she had made. So she created the French."

Polly Platt, in the hilariously practical book *French or Foe*, explains that much of foreigners' difficulty with French life comes from not knowing the proper code words or the importance of correct French social introductions.

Americans go into a bakery and order a loaf and expect the same service and politeness for themselves as for anyone else. Not the French. In France, if you do not want to own a burned loaf or simply be appraised by the bakery assistant as some undesirable kind of scamp, you come in, introduce yourself, and explain where you live and where you come from. Many first business meetings in France take place only so that such formal introductions can be made. After several hours of dining and wine, and no discussion of anything a foreigner considers even slightly pertinent or germane, perhaps you will prove to be someone monsieur can do business with.

Though younger people, habitually using Facebook and MySpace, have an easier time these days, many older adults tend to be a bit shy with introductions. You don't need to be one of them. I've met some of my best friends in airports, often waiting on interminable lines, because I noticed something curious or entertaining we possessed in common.

A social touch visit with coworkers or neighbors may produce nothing more than a sense of neighborliness or shared dismay. However, such contacts may sometimes prove to have significant uses. Expect the unexpected. Having a variety of contacts, allegiances, acquaintances, and colleagues can help you do more than get ahead. It may also help you survive.

DAY 18: SOCIAL REST TECHNIQUE 3— QUICK CONNECTS

Please take a look at lists 4 through 9 you made earlier:

4. people I like to talk to
5. people I wouldn't mind seeing more
6. people I can go to a movie with

7. people I can watch sports with

8. work colleagues I talk with easily

9. acquaintances I like

Choose one from list number 4, people you enjoy conversing with. When you can, call this person. But—make it someone who would not expect such a call.

You don't have to call on a landline phone. If it's someone you feel a little shy about and with whom you don't have a very strong connection, you can send an e-mail or text message. The script can go something like this:

> Hello, [person with whom you'd like to converse but probably does not expect your attempt at contact]. How are you? I just wanted you to know I was thinking about you, and wondering what you were doing. I've been doing _____ over the last few [days, weeks, years] and fondly remember _____.
>
> So how are things going for you?

This social rest technique is about making connections. For some it may be difficult, an action well outside their comfort zone. Practicing this form of social touch, however, often makes it easier for shy people to make social connections, even with strangers.

If it's still too difficult for you to initiate such a quick social connect, contact someone you are pretty certain will welcome such regard. Talk briefly and simply about daily things. Then return to your normal round of activities.

Quick connects do more than break up a day. They allow you to chart travels and communication to places you might not otherwise go, and they help you to create a fuller, broader social network

When and How to Do Quick Connects

The day of the week and time of day matter. For many, quick connects are best in the middle of an afternoon weekday when they feel a need for human warmth. Such quick connects often improve lagging attention in the otherwise draggy midafternoon, acting as a quick social reset button—if work conditions allow. Others prefer quick connects on weekends or evenings, when they expect the person they want to contact will be in and happy to take their call.

It's worthwhile to do quick connects frequently. Each week, or even better each day, you go down your list and see if you can connect with someone you would like to talk with. Some may be far away, some may be living within a mile. They may be people you knew well thirty years ago or relatives who knew you well during early years that you cannot yourself recall.

Quick connects can be very fast, and not just because the person contacted is busy or uninterested. There's something pleasant in knowing there is someone out there to converse with, often with a different spin on life than yourself.

These connections are restful because: (1) they can take you away from the hurly-burly you're often engaged in; (2) you converse, have human communication with someone you like; (3) they increase your social network, with the emotional, social, and economic benefits a larger network brings; (4) they help you recognize the much larger totality of which you are a part.

If you are writing by e-mail, you need spend no more than a minute letting someone know you are thinking about them. That type of psychological communication, particularly when its reach is broad and wide, may provide you with a quick sense of calm. Quick connects provide a fast way to get a stabilizing form of social rest.

DAY 19: SOCIAL REST TECHNIQUE 4—
WALKING TO LUNCH WITH A COLLEAGUE,
FRIEND, OR NEIGHBOR

That humans are walking machines has many benefits. Walking can help prevent heart disease, high blood pressure, diabetes, many cancers, and Alzheimer's disease. So why do so many people walk so little?

One major reason is cars. Few nations are as in love with or, depending on your point of view, as addicted to cars as Americans are. Rail lines within American towns and municipalities? Bought up by the oil companies many years ago. Fast trains between major American cities? A commuter line in a provincial Spanish city will get you to your destination faster than most American trains.

Neglecting walking means we have also neglected the pleasures of social walking. Spaniards may perform an evening stroll with the family or friends, but Americans, and now more and more Europeans, prefer to slink off to their preferred type of electronic entertainment or information machines.

It's too bad we've gotten out of the habit of walking. For lots of reasons, it's time to give it another try.

Walking to Lunch

Is there someone you want to visit but never seem to get the chance to see? Someone who has the best gossip or simply knows what your boss *really* wants? Do you have a neighbor who is attractive, strikingly intelligent, and much too seldom in your presence? If so, invite this person to lunch.

Pick a place where you can walk at least ten minutes in both directions, back and forth. You also want to choose a place that you know a bit about, though a little adventurousness can also go a long way in making the trip enjoyable.

One of the ironies of life is that you may need to drive to a parking space in order to walk to a restaurant. It's not an ideal solution, but it's okay. Just recognize that you need the exercise more than your vehicle does.

As you walk with your coworker or friend or neighbor, converse about anything that pleases you. Let your colleague do the talking at first, if you can. You do want to know more about this person, after all.

While walking, try to adjust your walking pace to the other person's. It may be slower or faster than yours, but that's fine. To make the pace easier, try to remember a piece of music that has a similar beat. Hear that music in your head, and try briefly to walk to that rhythm.

Notice the neighborhood. Just as you do with ear popping, look at the shapes, colors, and forms that you see. It's fun to pay attention to your own process of perception.

That includes the vibe you have while listening to your colleague or friend. Try to read their face. If you had to describe the emotions you see there, could you? Do they align with the emotions you're hearing in their words?

When you sit down to dine, see if you can tell this person a story from your own life that in some ways connects with what he or she has described. The story might be humorous if theirs was comic; it might have occurred in a similar place or at an emotionally similar time in your life. We often connect to others through our life narratives. Talk about one of your own.

On the walk back, discuss the environment you're moving through. Is it old? Undeveloped? Overdeveloped? Would you or anyone you know like to live there?

At the end of your walk, notice the person's reactions—physical, social, emotional—as you part. If it was a positive experience, try to set a date to do it again.

DAY 20: SOCIAL REST TECHNIQUES 5 AND 6— WALKING WITH A FRIEND IN A PARK OR THE WOODS, AND SEX AS SOCIAL REST

Physical activity is fully consistent with social, mental, and, as we'll see, spiritual rest. It's often enjoyable to obtain social rest in natural settings.

European research on this topic has generated more interest than what has been done in the United States. As noted before, studies done at England's Essex University found that people had far better moods walking through the woods than strolling through a mall.

But why? The researchers thought malls might bring out material cravings that cannot be filled. But walking through the woods connects people with something atavistic—the natural world we came from. Walking with someone else then allows us to share many wonders we might otherwise fail to notice.

There are multiple advantages to walking in woods or parks with sufficient greenery or beauty (beaches count too). Walking itself is physically healthy. Walking in natural scenery is, moreover, profoundly restful.

First, there is the effect of light. Light does more than reset and tune our biological rhythms. Light is a drug. Sunlight can be used to treat clinical depression. Particularly in the winter months, it can be used to prevent seasonal depression.

The numbers of people who respond to light are not small. Sebastian Kasper, now a professor in Vienna, did research in the 1980s that estimated a quarter to a half of Americans in the Northeast became depressed or down in the winter months.

Sunlight also activates our natural killer cells, very helpful in fighting viruses and perhaps cancer. It also helps create vitamin D. Many of us love the sun. We have good reasons to.

And there are multiple advantages of being in nature. We feel comfortable surrounded by greenery. The smells are different, the air tastes and feels different. In Britain and the Netherlands, walks in the woods are used to treat clinical depression as part of group therapy.

Increased movement also means increased blood flow. As you walk, more blood goes to the kidneys to filter wastes. More blood flies into the lungs and brain to bring nourishment and the materials needed to remake, rebuild, and grow new cells. Increased walking often lowers blood pressure, which can lower physiologic stress.

Now add to these benefits the many pleasures of social walking, conversing in a garden with someone you like.

How to Walk with a Friend in a Park or the Woods

Pick a spot that's mutually enjoyable. Walk with your friend or colleague for at least thirty minutes. Make a date for any time that works for both of you. It can be lunchtime at your job, a morning of a weekend day, or just after work on a spring or summer weekday.

As you walk, try to see what your walking partner sees. Talk about the environment around you and how it feels to both of you. Reminisce about gardens and walks you've had in the past. Recall that for many philosophers and sages in ancient times the most pleasurable experience in life was walking and conversing in a garden. No wonder a major group of Greek philosophers was named the Peripatetics.

Don't be surprised if while walking with a friend you start thinking new ideas or have a return of thoughts you have not experienced for many years. Thinking changes with place. New places are stimulating, just as they are restful.

If you can, do what English environmentalists do and create your own "green gym." As you walk through the park, pick up refuse to help clear walking paths. Make this an opportunity to clean the environment as well as sensibly use it.

When you're finished, thank your companion, and ask this person what she or he liked best about your walk. It might be the simple fact of social connection—a response you will certainly enjoy hearing.

In the future, it's good to visit different parks, beaches, and gardens. Yet if you can, go back sometime to that place you socially walked first, preferably with the same companion. Point out as you go what looks the same and what has changed since your last visit.

Consciousness is a process, just as growing, living, and resting are processes. Its nature is dynamic and shifting. Sometimes we enjoy things deeply just by noticing them, but it's often fun to share this knowledge.

And sometimes this sharing is deep.

Sex as Social Rest

This time, make sure the kids are away.

Most people have sex when they can, just before sleep. Though sex may aid sleep in some men, most couples' sleep is not improved by sex—at least in laboratory settings.

Sex works best when it's not a job. You don't want to make love according to a list of performance criteria. Sex is too powerful a form of social rest to sit around and measure it.

Something for a Sunday Afternoon

Outside of nurses, doctors, emergency workers, and tollbooth operators, Sunday afternoon is not when people generally expect to get lots of work done. Sunday afternoon then becomes an excellent time for creating social connection and productive rest. Late afternoon is also a time when male sperm are most motile—useful information if you're trying to become pregnant. If not, safe sex is always the order of the day. This Sunday afternoon you and your partner will find a sexy way to rest.

First, walk slowly together to the bedroom. As you come to the bed, lie down and face each other.

Talk about the moment you met. What did you imagine your

partner was like in that first glance? Can you remember his first comment, how you felt when she spoke?

Tell about your first sexual encounter together. Were parts of it funny? Try to remember your first kiss. What did you feel in your head, hands, arms? What might that experience feel like now?

Remember the best time you spent together, and tell your partner about that time. Tell it like it's a short story, with a beginning, a middle, an end. What did you love about that time, what embarrassed you, what could you not wait to feel again?

Now, touch each other. Do it quietly, slowly, like a sculptor who wants to appreciate every surface. Take your time. As you move, tell a story about yourself that makes you laugh.

Feel the hair, the eyebrows, the nose, the lips. Touch parts of the back and arms you might rarely notice. Feel the hollows of the skin, the furrows of muscle, the round swelling at thigh, belly, and buttocks.

Talk about what you really love about the other person, no matter how small or unimportant.

Forget about where you are, and think only of your partner. Feel them, their skin, their hair, the pulse of their arteries beneath your fingertips. As you relax, fantasize about what you would really like to do with them. If you know your lover well enough, go ahead and do it.

With luck, you can do more than make love. You'll feel loved, wanted, cared about, desired—feelings that have kept humans going for millions of years. That sense of inner rest and social and physical communion can also keep you going a long, long time.

Summary

Social rest is powerful. Social rest increases social connections, which can prevent heart attacks, strokes, and perhaps cancer; promote survival; provide personal meaning; and become simple, outrageous fun that will be remembered for a lifetime.

There are many, many techniques of social rest, but many involve social touch, the ability to engage in conversation with someone you know well or want to get to know better. In this chapter you learned how to.

make a special connection with someone you care about

visit a neighbor or coworker you'd like to know better

make quick, useful social connections

walk with a coworker to lunch

invite a friend to walk and converse with you in a park

use sex as social rest

Social connections are quick, slow, deep, broad, powerful, and amusing by turns. Knowing how to make even the most rudimentary social connections can make your life far more restful and enjoyable and allow you to develop relationships you may enjoy all your life.

chapter 6

SPIRITUAL REST

Spiritual rest can have profound effects on your psyche, your capabilities, and your relationships with others and the world.

The world is far greater than we know, but humans never cease trying to understand it. We never cease asking questions about our origins, our makeup, and our ultimate purpose. Some scientists argue that we are hardwired to be spiritual. Indeed, prayer and meditation physically change the brain.

You might try this quick quiz on spiritual rest:

1. Thought
   a. increases connections between nerve cells.
   b. may change nerve cell growth.
   c. can increase gray matter through many different parts of the brain.
   d. all of the above.

2. Spiritual rest markedly increases brain energy use.
   True   False

3. In brain imaging studies, deep meditation and prayer
   a. look very similar.
   b. appear to be completely different.

4. The benefits of religious practice include
   a. better mental health.

b. greater overall health.

c. decreased anxiety.

d. greater social connection.

e. all of the above.

5. You can experience spiritual rest

a. during the workday.

b. in your living room.

c. alone at home.

d. all of the above.

Answers: 1. d; 2. false; 3. a; 4. e; 5. d.

The brain almost always tries to seek order even in places where there appears to be none. Think about watching a movie. We see characters move across a screen in apparently real time. Their actions are continuous, fluid, and three dimensional. But is that what's really going on? No. In fact, what we are really watching is a bunch of photographic stills being projected on a screen one by one. Two-dimensional photos pushed before us in a continuous rhythm—that's it.

But we don't *see* a movie that way. If the photos are projected fast enough, above the physiologic threshold we call the flicker fusion rate, we visualize those images as continual and moving, a three-dimensional reflection of the world.

Our capacity to take bits of information and put them together into multiple ordered sequences can occur with tiny parts of images, fragments of music, or just the faintest touch. Our brains are built to create order, to see patterns. That is how we understand our world.

Yet most of our brain function is really self-reflective. Neuroscientist Marcus Raichle, writing in "The Brain's Dark Energy" for *Science* magazine, reflected on what the brain normally does. He used the metaphor of dark energy because so much of what the brain does is unknown. Raichle figures about 60 to 80 percent of total brain energy consumption is spent communicating between individual neurons and their support cells.

What happens with major stress? How much of a load does a crisis or difficult task add to the energy output of the brain? Perhaps .5 to 1 percent. *Most of what we do does not change overall brain energy use.* It seems that most of the time and energy the brain uses is spent "talking" to itself.

That's a lot of energy. The brain normally takes up about 20 to 25 percent of total body fuel consumption.

A large part of this energy is engaged in rebuilding, rewiring, and renewing the brain. That also automatically means rebuilding the rest of the body. The brain is the executive organ. Almost everything in your body communicates with the brain all day and night, except during some periods of sleep. Well below our level of consciousness, at almost every moment, the brain is helping all the other organs redevelop and rebuild themselves.

Rest, the restorative, renewing activity of our bodies, takes up a lot of time and work. Raichle was also the namer of the brain's "default mode," the electrical and blood flow patterns seen when the brain is passively resting. This passive rest is in fact very active. In a paper published in 2007, researchers A. M. Morcom and P. C. F. Fletcher pointed out that though the differences are small, in some of the brain's rest states more energy is used up than when the brain is performing set tasks. There's a tremendous amount going on in default mode, even if we don't quite know what most of it is.

Clearly lots of brain activity is spent in rest, just as occurs during the different states of consciousness we call sleep. Aware or not of what our brain is doing, we are spending much of our time thinking. And that thinking shapes and redirects brain chemical and electrical activity. It also changes our brain's anatomy.

Thought Is Action

People have a hard time imagining that thinking changes the physical structure of a brain, perhaps because they can't see that process. Yet that is precisely what activity does everywhere in the body—it physically changes the parts of the body being exercised.

We can see this easily when we look at our voluntary muscles. If I sit around in the bed or stare all day at a TV set, my muscles will not grow. People know that when they're in an accident and stuck for weeks immobilized in a cast, their muscles shrink.

But people also know that if they start pumping iron, and keep adding on weights and doing repetitions, almost any beginner will become stronger. Most people can see their strength grow through added muscle tissue. Arnold Schwarzenegger did not get to look the way he does because he was born that way. (Bigger muscles are easier to produce in males, I'm afraid. Most women and some men lack the genes that make stronger muscles look a lot bigger.)

Muscles change with use. Just like everything else. That includes the brain. If you use parts of the brain long enough and repetitively enough, they get bigger. Brain activity leads to anatomic changes.

One way to research this is by studying people doing meditation. Meditation has been a large part of the world's spiritual traditions for thousands of years, though it is historically practiced more in Asia than in the West. However, use of meditation techniques is growing throughout the United States and Europe as health practitioners are discovering its many benefits.

Meditators grow their brain. You can't see the change without MRI scans, of course, but different parts of meditators' brains grow larger.

Long-term meditators grow bigger, fatter frontal lobes. Frontal lobes are where we concentrate, fix our attention, plan, focus, and do much of our analysis of problems.

Even novice meditators grow bigger frontal lobes, with fatter sections of gray matter. One study done by Andrew Newberg of the University of

Pennsylvania, who studies the effects of religious practices on the brain, showed that poor recall could be improved by meditation training.

Meditators grow more brain tissue in other brain regions as well. European research demonstrates meditators building up more gray matter in the midbrain, which handles housekeeping functions like breathing and blood circulation. If you practice active rest techniques, you may soon experience such changes as well, even if you don't get an MRI scan to find out what they look like. You too should be able to grow parts of your midbrain by practicing physical rest techniques like deep breathing and mental rest techniques like UnNap naps.

Meditators also grow more gray matter in places like the dorsolateral prefrontal cortex, important for muscle coordination and active memory, and experience changes in the structure of the thalamus, a part of the brain critical for processing information flow from all parts of the body.

So there is a lot of evidence that trained conscious thought, like meditation, changes the brain physically and chemically. Meditation, much like prayer, also changes the thinker.

Most meditation traditions use techniques that lead people to lose or entirely undo the distinction between themselves and the world around them. Our overpowering sense of a separate self—of individual uniqueness, of a separate, thinking brain constantly struggling with the world—loses its hold on us when these techniques are skillfully performed. Meditators often state they have experiences in which they feel no separation between themselves and the physical universe. Some talk of the "false I," the insistent voice in our heads that is the expression of our normal individual consciousness. While meditating, they instead feel their subjective and objective worlds dissolve, that their consciousness unifies and becomes one.

This dispersal of a sense of self is felt by far more than meditators. Many who pray daily speak of experiencing an "oceanic feeling," a sense of unity with all that exists. Nature is high, wide, vast, and deep.

The Matter of Matter

Just as the brain automatically tries to establish order and pattern, much of the edifice of human science does the same. Physicists, who try to create and describe models that fit the physical world, are finding out that the nature of nature is both weirder and more awe inspiring than we used to think.

Recent mapping of our galaxy and other galaxies is showing that things are not what they seemed. Normal matter and energy, the stuff we live in, is presently thought to constitute perhaps 4 percent of the stuff of the universe.

That's *one twenty-fifth* of the known universe. What is in the other ninety-six percent?

The physicists' answer—dark energy and dark matter. The vast majority of our physical universe is made up of stuff we know hardly anything about. Raichle's metaphor of dark energy extends well beyond the brain to the rather large universe our brains spend so much time trying to figure out.

Physicists have spent much of the last several decades trying to join quantum mechanics and relativity into the Grand Unified Theory that eluded and frustrated Einstein much of his life. Some popular GUT variants use versions of string theory, which (very roughly) sees the universe as made of an infinite series of almost infinitely small "strings."

Trying to think of these strings as "matter," such as baseballs or sofas, is a major stretch. Most string theories need eleven to fourteen dimensions or more to model our tiny three-dimensional universe (four, if you count time as a separate dimension).

Some people, like me, have trouble visualizing from two-dimensional architectural drawings how those plans will look in three dimensions. Now, if you can, extend that planning function to eleven or, better, fourteen dimensions.

It's no wonder many cosmologists and physics theorists think it likely that our universe, which includes infinite numbers of galaxies like the rather pedestrian one we inhabit, is itself a single instance within an infinite number of other universes. Indeed, according to some theories, every possible universe has existed, is existing, or will exist, just as everything we know or experience has happened or will happen. Such images allow some scientists to contemplate our world as nothing more than a simulation, a computational matrix. For them, our galaxy and universe may exist as a kind of vast computer game, just as in science fiction movies like *The Matrix*.

With a natural universe this marvelous and strange, there is much room for spiritual contemplation. The many ways this can be done all magnify the health and psychological benefits of spiritual rest.

The Power of Prayer

Religion powerfully affects health. In various American studies, people who regularly attend religious services live longer than those who do not.

Much of that advantage may lie in social connection. People who attend religious services are part of communities that deal with more than the spiritual. People educate each other, help each other out, give to each other. According to University of Michigan researcher Neal Krause, there is evidence that those who provide help to fellow parishioners fare better healthwise than those whom they aid. Indeed, to give is to receive.

These religious connections are also more than social connections. Parishioners engage in social community and a fellowship of religious belief. Being a believer itself has definable health benefits. Those who experience some sense of meaning in the world appear happier and healthier than those who do not.

The health benefits of directed prayer are controversial. Studies done in the nineties showed that prayer can help the health of those prayed for. Recent studies have not duplicated these results, particularly if the people being prayed for don't know they are being prayed for. However, for many, prayer is one of the most powerful experiences they know. Even if they can't tell you why, they feel better after prayer.

Andrew Newberg of the University of Pennsylvania has worked extensively to image what happens to the brain during prayer. Much of what he has found is remarkably similar to what happens with meditators.

When people who are praying feel physically separated from the earth, Newberg found that the brain's laterally placed lobes, which help define position sense, lowered their activity. And the way someone prays certainly affects the results. One study by Newberg looked at levels of anxiety among religious followers. Catholics who prayed fingering their rosary beads felt less anxious than those who said the Lord's Prayer. Fingering the beads appeared to make them calmer. Among Zen monks in Japan, though Buddhist meditation quickly leads to mental relaxation, very different parts of the brain are turned on or off by concentrating on different sutras.

Yet for those who do not believe in God or religion, prayer may evoke feelings of hope, wonder, and connection with powers greater and wider than themselves (yes, people who do not believe in God still pray to forces they cannot define). Often such prayers are accompanied by physiologic changes that presage personal restoration and a sense of psychological renewal. Techniques of spiritual rest may have powerful results for nearly anyone. Contemplating the nature of nature, the width and forms of our universe, can become highly restful. Our brains are built in a certain way. We do appear hardwired to attain spiritual rest.

Here are a few spiritual rest techniques that most anyone can quickly practice.

DAY 21: SPIRITUAL REST TECHNIQUES 1, 2, AND 3—PRAYING FOR ONE MINUTE, MOVING THROUGH TIME, AND MOVING THROUGH SPACE

For sixty seconds, pray for a good result. Focus your mind on a better future for someone you love, or a good outcome for someone you do not even know. You can pray for a change in the state of the entire world or for the smallest act that will help anything or anyone you care about.

You can pray to the God you believe in. You can pray to the powers you think exist in the universe but that you have never seen demonstrated and are not even sure of. You can pray to a spirit, a divinity, or for a sense that something good will issue from your thoughts.

You may pray knowing how deeply you are connected to others, in that you know there are those who think as you do and desire the same ends. You may pray in the hope that others will come to feel about the world as you do though you have no sure knowledge of that and indeed possess only scant belief. You may pray in the assured belief that someone will hear your prayer or in the uncertain belief that no thing or entity you can grasp will ever acknowledge such clearly expressed requests.

You can pray at the start of the day upon waking or in the last glimmerings of consciousness before sleep. You can pray on a Sunday or a Monday or a Wednesday, in the morning or the afternoon or at the end of twilight.

You can pray while you are in the car. You can pray while waiting in a lunch line or walking through a park. You can pray in a Taoist temple or a Catholic refuge, in a synagogue or in a mosque. You can pray wherever you are as long as you concentrate on hope, on fellowship, on the prospect of aid for any that you think worthy of consideration.

Just pray for one minute. Pray as you are, and pray as you hope to be. And when you are finished praying, think about yourself. Is your body calm and feeling at rest? Are you feeling a little more hopeful, just a slight bit more joyous? Do you feel that you have connected a little more with others—others who may sense the same needs for the world that you do?

Regardless of the outcome, pray. The way our bodies and brains are designed and configured, the act itself may prove enough. Prayer and the attempt at spiritual connection are also forms of giving. In that way they are their own reward.

Moving Through Time and Space

Spiritual rest connects us to things larger than ourselves. It's useful to sense how large space and time are. Contemplating the vastness of either time or space can provoke in us a feeling of awe, and with that awe comes a great sense of rest.

Moving Through Time

The technique of moving through time demands a spatial starting point. You want to choose a stable or stationary place, preferably at or near ground level, like the desk or dining table where you're reading this, whether it is in Brooklyn, Omaha, or Bangkok. Starting to use this technique for the first time while flying at thirty-five thousand feet or riding on a subway or commuter car may make the process difficult.

Begin by visualizing the place where you are sitting. Now imagine it exactly one year ago. The desk or chairs may have been in precisely the same places they are now.

Or perhaps someone else was inhabiting this room. That person may have configured it in a completely different manner. Furniture, colors, the whole framing of the space may have been entirely different.

Now as you sit, reach back in time. First go back ten years. The room you're in probably did look different. The electronic devices were probably less powerful and bulkier. New paint may have lined the walls. Chances are pretty good someone else was living there. The person in this room may not have been from your family and maybe was not even someone you know. From whatever information you possess, try to imagine how the room looked.

Now reach back a hundred years. There may not have been any buildings where you now sit. Where I live in Florida, all that would have been seen of my present living space was a shell mound. No condominiums stood towering over a placid sea. A few houses strung out in an irregular necklace faced a bay and a small pier, with most of the then-meager population living farther north.

Go back a thousand years ago, and there were probably far fewer people about. The people who lived in my neighborhood, the Calusa Indians, are extinct. Artifacts from their lives are rare, their culture and beliefs and spiritual yearnings a matter of guesswork. If you are sitting or now standing somewhere in North America, chances are good that no permanent settlement existed a thousand years before in the same place. People who lived and died in the land where you now sit probably left no written records, indeed few records of any kind.

Ten thousand years ago, the chances of human habitation where you presently sit are yet smaller indeed. Where I write this sentence, there may not even have been land. The majority of Florida is a sandbar. Sand shifts. Where I sit, manatees may have amiably drifted by, scooping up vegetation while dolphins swam free in multiple small coves.

Let's skip further periods of time. A million years ago, chances are good the land on which I live was ocean. Odds are high it teemed with wildlife, stuffed with many species now lost to us.

A hundred million years ago our world was a place we might not recognize from space. There were different continents, different life forms, sufficiently strange as to appear extraterrestrial to our present eyes.

Go back a billion years, and where you sit was probably rock or molten lava. If there was life around you, it was small and hardy, daily facing conditions that might kill almost everything that we today think of as alive.

Go back a few more billion years, and you will see nothing. There is no earth, no sun. Instead there is the emptiness we call space.

Not that it was ever truly empty. Dust, perhaps even rocks, flew around at speeds that make our lives seem frozen in place by comparison. And the vastness of this vacuum may have been more appearance than reality. Throughout the depths of space, fields of energy existed that create the conditions for both what we call gravity and what we call matter. If there was matter present in this vicinity at that time, most probably there was also dark matter. If indeed there were no viable forms of energy we presently recognize, there should have been dark energy, creating the edifice for the particles that today make up people, planets, and stars.

It may be impossible to imagine thirteen billion years ago. To a child a morning is a long time, so what is the measure of half a trillion mornings?

Still, try to imagine that time. At this point the universe does not exist. There is true emptiness. There is nothing, far less than most minds can imagine as nothing.

Suddenly something happens. A universe is created. What would you have seen had you been there? Who else can you imagine present?

Now move forward again in time, using these images of the different periods. If possible, modify them, making your own images. To make it easier, see the place where you now sit as a series of changing still photographs, one for each epoch of time.

See all those images, one by one, separately and together. When you finish, visualize yourself here and now. Think of where you now sit, reading. Look at your watch. You may have spent several minutes doing your time travel, but chances are you covered all those epochs in less than a hundred fifty seconds. When performing spiritual rest techniques, it can take only seconds to advance yourself billions of years.

Moving Through Space

The Chinese like to talk about the great within the small. Often they see the small as, on its own terms, equally as important as that which is large.

As you sit reading this sentence, go back into your body and search out your heart. Visualize your beating heart.

It's a fairly small organ, not much larger than a fist. But it does a mighty job.

Now visualize the left side of your heart. The left ventricle is the pumping station of the body. It forces oxygenated blood through the large arteries that feed both the heart and the rest of you, all but the lungs.

Try to image one of those arteries supplying the heart. Its diameter is thinner than the width of a paper clip, but it's critical to your survival.

Follow that little coronary artery in the heart. Soon it will begin to branch, then branch again and again. Quickly the branches twist into tiny arterioles that feed a small wad of heart tissue. Inside that bunch of tissue, pulled out like pearls on a malformed string, stands a necklace of tiny electrical generators.

These are pacemaker cells. They are the origin of the rhythm of your heartbeat. Shut them down, even mildly derange their cycle, and everything you think about may stop, including you.

Right now your pacemaker cells are strumming along, banging out their rhythms in remarkably regular music, thanks to the influx and outflow of ions across their cell membranes. Those cell membranes are filled with cholesterol, the stuff you brag about lowering through modifying your diet, beginning to exercise, or swallowing a statin pill each night. The cholesterol molecules lining your cells' membranes are fungible, movable, and fast.

Through those cell membranes innumerable molecules come and go: proteins, glycoproteins, large slimy fats, and tiny ions. These membranes are the borders of your cells and the means of communication, the information boundary points, and the engines of chemical transmission. Everything they do is critical to life.

Let's go a level smaller. The cholesterol molecules that line your pacemaker cell are made up of carbon, hydrogen, and oxygen atoms, configured for action and transport. Choose one of those carbon atoms.

At the center of your carbon atom lies its nucleus of six protons and (usually) six neutrons. The rest is, like the vacuum around the stars, mainly "empty" space.

Except it's not quite empty. Electrons fly around the nucleus, making shells of orbits that spread all the way from your heart to the center of the moon.

How can these electrons orbit your cholesterol carbon atom and the lunar surface? This is all a matter of probabilities, not absolutes.

Quantum mechanics specifies that the placement of matter is never a sure thing, including the probability that one or more of your electrons briefly moves past the moon.

What's inside the protons and neutrons? Many other particles. Some are quarks, incredibly tiny particles that can only be inferred by experiment. Quarks themselves may be composed of hidden, curled pieces of vibrating strings, matter as abstract as the nature of infinite infinities.

Now it's our chance to move back up the spatial scale. Let your mind visualize the different levels of magnification. Start with wispy, extraordinarily small subatomic strings or particles. Then progress to atoms, with their nuclei surrounded by orbiting electrons, then to molecules, like the cholesterol embedded in your cell membranes, and finally visualize a heart cell. Though we have moved up through many different layers of space, we are still thinking of objects too small for our eyes to see.

So let's look around us. We are probably sitting in a room, in a building possessing several other rooms. Many other buildings may surround us. Some of these edifices may reach hundreds of meters into the sky, but the sky dwarfs all human constructions.

Our building lies in a town within a state or province of a country, just one of the nearly two hundred nations on the Earth. Above us lies an atmosphere perhaps a hundred kilometers thick, whose composition ranges from dense, foggy cloud banks near sea level to sparse air molecules, to wisps, to no air at all.

Our planet is a small rocky place. Estimates are there are one billion planets like it within our own galaxy, which compared to those surrounding it is small and unmajestic.

All that we know and see, the stars at night, the nebulae beyond, are little ripples in a rip that began flying outward thirteen billion years ago.

Now you can return back to where you are. Move through space, travel those thousands and billions of light-years, and go back to where you sit. It's a long journey, a worthy journey. Perhaps what's best is that to make this journey, you need not leave the spot where you sit looking at it all.

How and When to Practice Moving Through Time and Space

Like many spiritual techniques, moving through time and space takes a bit of concentration. The wonder is how little you need to go long and far.

These techniques work best when you have the time or the need to seek perspective. In times of social or economic crisis, you may want to use spiritual rest techniques regularly.

Try to do spiritual rest techniques when you do not feel rushed. With practice, moving through time and space can each be done within a minute, if you watch the time and spatial shifts as a series of mental slides or as a film. Of course, you may want to take more time for these techniques. Doing spiritual rest techniques for three to five minutes in a place where you will not be disturbed can prove highly effective.

Each opportunity you have to use these spiritual rest techniques may be customized. You can do them in a neighbor's home or on a picnic in a park. Often they are most enjoyable to use when you are traveling or visiting new places.

When you practice moving through space and time, you can imagine the dinosaurs flying through the great swamps of what will become North America or see the eruptions of slag, sulfur, and mud that created the billions-of-years-old life forms discovered only in the last few decades. Our lives follow time's singly pointed arrow, going only forward. This is not true for our imaginations or our physics. Quantum mechanics works equally well moving both forward and backward in time, creating a beautiful mathematics that allows us to better sense what created the galaxy around us.

When you can, take the time to take time's measure by practicing these spiritual rest techniques. The Romans were right—time rules life. Time is also much larger than life. If Einstein is correct, so is space.

DAY 22: SPIRITUAL REST TECHNIQUE 4— CONTEMPLATING SUCHNESS

Zen Buddhist teachings are subtle and encompassing, but even crude approximations of their techniques can allow nonbelievers the quick benefits of spiritual rest. With a little practice, contemplating such things can deliver to you a combined sense of awe and peace.

Classical Zen teaching describes a universe of samsara and nirvana, the world of sensation or what we call subjective reality, and the real world that is. Our world, where we eat, sleep, strive, and love, is for them a place of pure illusion.

Zen meditation demands detachment, an ability to anchor ourselves outside the world where we live, laugh, and worry. Zen masters, through guidance and practice, then teach others to detach themselves from this detachment.

You don't have to become a Zen master or a Zen acolyte to truly enjoy the beginning of this process—contemplating suchness, all the world where we live.

Now that you've learned how to go forward and backward in time and space, find a comfortable place to sit and think.

Sit straight. Breathe deeply, in to the count of four, out to the count of eight. Feel your breath move across your lips. Then hear it.

Now look around you. Start by naming everything you can see— pictures and prints, bookshelves and linoleum, books and papers and pins, cell phones and floorboards and rugs.

Next, use your mind's eye to travel outside the room. Visualize the room you're in as part of a building. Next, visualize the entire building you're presently in.

Extend your mind's eye. Your building is probably part of a neighborhood, a set of other buildings usually filled with people. These neighborhoods combine and form cities.

Spend a few seconds thinking of the size of a city and all the people there. Think how many lifetimes it would take to meet them all.

Yet a city is itself usually only a small part of a nation. Nations themselves are often parts of civilizations that are centuries or millennia old.

Think a moment of the history of your civilization—all the people who lived and worked and strived and hoped. Imagine all that they created. Next, consider what you and your friends hope to create.

Now it is useful to move past the strictly human world, which takes up so much of our attention. Consider all the plants around you—the grass, shrubs, and trees that live in your town, your province or state. Through a mental leap, try to imagine seeing all the plants that live on the Earth.

Consider now our planet's animals, the many millions of different species. Now search with your mind's eye, and go below the surface of the Earth. That's where the large majority of life on Earth actually lives.

You don't see them, but they're there. Far more than moles and worms live beneath the surface. The largest single biomass of our planet is made up of the nearly uncountable bacteria and "simple" life forms that live below ground. Invisible to our eyes, they remain necessary to our survival, forming the soils that allow us and nearly everything else to survive.

Consider all that life. There are one hundred trillion organisms in or around you alone, and near-infinite amounts everywhere else. Most are too small to see. Others, like whale sharks and blue whales, are the size of buildings, and each is connected to the others by chains of energy and information that allow all the others to live.

For the next moment, go and visit in your mind's eye the nonliving parts of the world—the seas and lakes, the mountains and hills. Below the oceans and the giant ridges of rock that form land lie tectonic plates that move the continents around like mammoth bumper cars. Beneath these plates lie molten shells of immense dimension. At the center of the world flow rivers of metals that arrived there from the plasma cores of bursting stars. Much of the matter sitting at the center of our planet was shot out from exploding supernovae billions of years ago.

Contemplating suchness lets you see all these things, individually separate and connected. You can sense their forms and shapes, their elasticity and power, within mere seconds of mental time. Our world of suchness is infinite, but also great is your ability to see, hear, feel, and imagine all that is present.

Contemplating suchness can give you feelings of vastness, greatness, wonder, and awe. These are restful feelings, feelings of appreciation. At the beginning, try to contemplate suchness for at least five minutes. Later, with practice, you may contemplate the world around you far more quickly. After a few weeks, elements of suchness can be contemplated within the time spanned by the few blinks of an eye. Yet no matter how much you appreciate, there will always be more than meets your inner eye.

DAY 23: SPIRITUAL REST TECHNIQUE 5— SEPARATING SUBJECTIVE AND OBJECTIVE

Meditation is practiced for many reasons. It can connect you with the infinite. It can make you recognize you are part of something much larger. It can start you down the path of placing your consciousness more and more under your control. Even without these powerful abilities, simple meditation can provoke a profound sense of rest and, with that, a delicious feeling of freedom.

Most people don't feel they have enough time to meditate a half hour or an hour each day. Many change their mind when they see how simple meditation can be to learn and use. Simple forms of meditation can rapidly calm and rest the mind and body. Here are two simple techniques, which are really versions of the same thing.

Version 1: Think Like a Fly

It's hard to get away from yourself. Most times, we don't even think of making the attempt. It's hard to imagine separating "I" from myself.

Meditative practice can start to dissolve that barrier. You begin with something small but lively.

A fly.

Most of us don't consider flies to have the benefits of selfhood. We don't worry about flies' personalities or their unique individuality.

We feel different about dogs and cats. To us each of them seems unique. Though theologians argue about whether dogs and cats have souls, we feel that they are very much individuals. They see us, look at us, watch us, emotionally respond to us.

We don't see flies in the same way. If we feel one buzzing nearby, we sometimes bat our hands around. Then they probably fly away, a good survival strategy if you're a fly.

To start this short meditative technique, you want to get away from thinking, breathing, moving as a unique human individual. You want to briefly imagine yourself as a fly.

You don't worry. You don't think. Your life is relatively clear-cut. You see. You sort of hear. You move. You smell.

Mostly you react. If you smell things that perhaps mean food, you move toward them. If you smell things that may be dangerous, you move away.

If there's a great deal of light, you may move quickly—there is simply too much danger present. If there's just a little light, plus the right kinds of smell, you move forward.

You're not thinking as you, a separate person. You're not self-consciously contemplating or considering or analyzing.

You just do. You move. You fly. If the air feels right, you land on a surface. You touch with your feet, stick your tongue onto it, consider if it's okay or not okay to stay.

If you find food, you stay. If you don't, you may still stay. You can stand and wait for the smells of food, changes in temperature and light, or the presence of other flies.

As you imagine yourself roaming the world like a fly, don't think of poor Jeff Goldblum in David Cronenberg's movie *The Fly*, part human and part fly. Instead, imagine yourself roaming just as a fly, living in that world of sensation and action and split-second decisions. Here is a world without much past or future. Things move, and things are. In the next second, they move again. Existence and the urgencies of action are all there is.

Once you've been able to imagine yourself as a fly, stay in this world for one to two minutes. Don't be surprised if your eyes are a bit sharper, your mind a bit fuller, as you return to the world where you normally live.

The fly's world is both much smaller and much larger than your own. It's larger because it includes everything. The fly is not alone but part of the whole.

Version 2: Simple Observational Meditation

This technique is used to relax the mind and center you in the world. At the start it will take you three to five minutes, though it can be performed for far longer periods.

Find a comfortable spot, and sit. Breathe in slowly, feeling your abdomen expand and contract. Feel the air move cross your lips, flowing down your throat, traveling into the dark, lively spaces at the bottom of the lungs.

As you sit, observe the thoughts that come through your head. Hear them, their words. See the images, and name them.

There's this paper you have to write . . . the birthday present for your mother . . . that wallpaper is a really funny color . . . darn, my hand still hurts a little from that scratch yesterday morning . . . was that marijuana I smelled outside the post office yesterday? . . . I love pizza, even if it's bad for me . . . is this random flow of images and ideas how I really think?

Hear your thoughts, see them, better yet, feel them. For a short while view yourself with a little detachment. You are the observer, observing yourself. Watch your ideas and images, your sensations and feelings, just as if you're watching birds passing across the sky.

Or studying a fly sitting on your desk.

As you observe, sit back a little. Ask yourself these questions: What would an observer observing me see? What if I had a second self inside me that just watched what I did?

Maybe that would be fun. Whatever the thoughts, go with them, as you would while reading a pleasant book. Just feel, let the thoughts and images pour through your mind.

You can use the technique of separating subjective and objective whenever you have a minute or two and have a comfortable, safe place to sit or stand. Meditation demands concentration.

Yet the rewards can be both fast and lasting. Seeing the world with different eyes is restful and restorative. It also lets you see things you don't normally look at.

You become more alert, aware, and alive—some of the real benefits of rest.

Summary

The world is much larger than we know. Spiritual rest connects us with things larger than ourselves, in ways that seem hardwired into our brains.

The human brain loves patterns, but spiritual rest allows us to detect and enjoy patterns that may appear beyond our comprehension. Such techniques can provide a profound sense of rest, of inner peace, of connection and self-healing. They also change more than our brain chemistry; they change the very anatomy of the brain itself.

In this chapter you learned how to

pray for a minute

move through time

move through space

contemplate suchness

separate subjective and objective

Spiritual rest has been used by people for thousands of years. Shrouded in mystery, it is actually easy and simple to do.

Simple things work. The body is built to rest—physically, socially, mentally, and spiritually. Many things lie within our grasp when we know how to rest.

chapter 7

REST AT HOME

Resting at home has many advantages over other places where we rest. They include the following:

1. You can often put together the different kinds of rest— physical, mental, social, and spiritual—more easily at home than anywhere else and let them work together synergistically.

2. You can rest socially with greater ease. There are many advantages in resting with people you love and care about.

3. Usually you will have safer and more comfortable places to perform rest techniques at home than in other places.

4. You generally have more time and control of your types of rest at home than elsewhere.

5. Different techniques, like those involving self-hypnosis and spiritual rest, can be performed at home with greater ease and comfort than while at work or in public.

6. Others can help you rest and can collaboratively create with you new ways of resting that include putting different rest techniques together to fit the rhythms of your life.

There are advantages to resting together as a family as you build your social connections during group rest activities. Yet families are made up of individuals, and different generations have rather different attitudes toward when and how they should rest. We'll now address these differences so as to optimize resting at home. Many of the differences between individuals and generations affecting how and when we rest are biological, changing with age and through the varied individual internal clocks that time our lives. Beyond our many individual differences, our biological clocks also change through the life cycle.

Rest and the Life Cycle: Youth

Adolescents don't like to rest very much. The sad part is they need rest more than other age groups.

They certainly can learn to rest. However, it takes a little education on the part of both teenagers and parents.

First, adolescents have to recognize how much they are growing. Not just up, out, and too often sideways, but inside their brains. In the 1970s, Irwin Feinberg showed that 30 to 40 percent of the synaptic connections in the brain die during puberty.

Synapses are the business region of the nervous system. That's where much of the most important communication between nerve cells takes place. The basic anatomy of the brain utterly changes during the teenage years—even well into the early twenties.

So adolescents are not just rebuilding and restoring during rest, they are also building and creating major new structures in their brains. That's a lot of work. It takes time, and it takes a lot of time for rest. Unfortunately, teenagers don't want to take the time they need to grow and redo their brains.

They especially don't want to take the time to sleep. To truly grow the brain to think and learn well, adolescents need about nine to nine and a half hours of sleep each night. According to the work of Mary Carskadon and her coworkers, lots of teenagers are getting seven hours of sleep or less each night. Many, way too many, are getting six hours of sleep or less most nights.

You see the results in multiple ways: (1) adolescents sleeping through the first half of classes; (2) kids getting lower grades; (3) bigger, more obese kids; (4) cranky, tired kids. Getting adolescents to sleep sensibly can completely change all that as well as improve family relations.

Yet it's hard to convince kids they need to sleep in order to think effectively, learn more, and do better in school. And that's just the sleep element of effective rest. Who wants to rest at all when there are so many more interesting things to do? How can rest compare with video games? Movies? Parties? Hanging at the mall? Instant messaging?

It's simple—you explain to them how cool rest is.

Take, for example, hanging out at the mall. If they go to the mall with their friends, it can provide a form of social rest. Social rest is lots of fun when you talk together, and teenagers can walk to music (mental rest technique 3). Many kids can walk to music while moving with their iPods, which allows them to choose their own music *and* mix and match their favorites with the tunes brought by their friends.

Teenagers also love games. An easy game to play is to walk to music with one person listening to a tune—and then have the others figure out what tune it is. Some people do more than walk with the rhythm of a tune; they try to embody it with their movements, which often gives their friends clues about what they are listening to. Then everyone has the chance to get in the rhythm.

And if teenagers are absolutely set on playing a video game, have them train first with self-hypnosis and focusing the eye (mental rest techniques 1 and 2). Both techniques aid concentration and performance. Teens can use the same quick mental rest techniques before working on term papers or preparing for an exam.

Movies are fun when you see them with friends, but they're even more fun when you discuss them afterward. Some teenagers may be a little confused as to what the director was trying to do, but they can wait and ask friends, who can provide different opinions. And what was it that made the acting so credible, so real? To find out, you can have them call a good friend they haven't seen for a while, making a special brief connection (social rest technique 3), to learn what they think. Even better, you can ask your kid to join you in conversation about the movie while walking in a garden or park (social rest technique 5).

When parties get too fast, too strange, or too crazy, teenagers can rest up and reset very rapidly using ear popping (mental rest technique 4). Once they have heard their ears pop, carefully looked around, and recalibrated their perceptions and their sense of place, teenagers are in better shape to figure out whether they will want to stay where they are. And if they decide they do want to stay, they've had a chance to check their own plans and perhaps consider new ones. Maybe they will want to go over and try some new steps with that fantastic dancer over there in the corner, the beginning of what can become a lovely, pleasant flow experience.

Instant messaging often takes on a life of its own, and as Gaby Bader of the University of Gothenburg has shown, dozens or hundreds of daily instant messages greatly interfere with nighttime sleep—and pretty much everything else. Yet kids experience an almost physical rush when they plan and then try to reach true friends, or just people they would like to become friends with, as can be done with quick connects (social rest technique 3).

The biggest problem in getting adolescents to rest remains convincing them to take enough time for sleep. You can argue with them that their grades will probably get better. That will do the job with some, though not as many as it should. You can tell them getting enough rest will make them feel more alert and less tired each day, but many teenagers don't know what to do with the energy they already have.

Fortunately, there are ways to convince adolescents to get sufficient sleep. They care a lot about some things, like their appearance and acting cool. Getting rest improves both.

Get too little sleep and you will get fat? Most adolescents have never heard such a bizarre fact. They really can't believe it. If I'm up all night and using all that energy, how could I ever, ever get fat? Yet show them the evidence, and they start to think a little differently. Most adolescents, including the skinniest, don't want to get fat.

And unlike the pencil-thin arms and legs of previous generations' heroin-chic female fashion models, being fit, trim, and muscular is once again considered an attractive look for girls. Sleep is when you produce growth hormone. Growth hormone builds muscle and sinew. You need enough sleep to get growth hormone, which means more sleep can help you look good. Boys' concern for their appearance usually means they want fit, powerful bodies, and proper sleep will help them appear fit and strong.

If adolescents still will not take enough time for sleep, you appeal to their simplest form of vanity—their skin. The skin on your face regrows in less than two weeks. A lot of that epidermal renewal takes place during sleep. Kids should know that in many ways, sleep literally is beauty sleep.

Rest and the Life Cycle: The Elderly

Elderly people, unlike adolescents, are not opposed to rest. They know they need it. They feel it in their bones.

One of the problems of aging is that our normal rebuilding and restorative processes shift as we grow older. Most athletes see slower times and weakening muscles once they pass the age of sixty no matter how hard they work out. The normal osteoarthritis that occurs with age is direct evidence that our restoration of joints and ligaments is not as effective as when we were young.

The problem for the elderly is not whether they need or want rest but how they view rest. For cultural reasons, many older folks see taking a rest, like a nap in the afternoon, as a feeble act, a version of laziness.

Part of the trick of aging well is to recognize that the body does not restore itself as quickly or as efficiently as we age, so you need to allocate *more* rest time to accomplish the same rebuilding. Once older folks recognize that the body, even when they are past a hundred years of age, is still replenishing and rebuilding itself, the prospect of rest loses this sting of laziness and idleness.

However, rest changes with aging for another reason. The biological clocks that determine alertness and sleepiness—when we do things best and when we do them worse—also change. Over the years from age twenty to age seventy, the average individual body clock usually moves forward, getting earlier by about ninety minutes.

As we age, we want to go to bed and get up earlier. It's not laziness that makes Grandpa go to bed at 9:00 p.m.; it's the force of human design. Human beings are built on time.

Rest and Body Clocks

Everything we do is affected by our body clocks, especially our twenty-four-hour body clocks. As you have already learned, our levels of alertness vary with our internal temperature rhythm.

A rising body core temperature means increased alertness, while a decreasing one makes us slow. If your body core temperature goes down quickly, as it usually does at night, we get real sleepy. When the curve of the body core temperature is flat, as in the early to midafternoon, we can easily take a nap.

Body clocks vary a lot from person to person. They also vary *within families*. That's important if you want to rest as a family.

Perhaps 70 percent of the population are sparrows, possessing average body clocks. These are the people who have no trouble going to nine-to-five jobs and sacking out by 10:30 or 11:30 p.m.

However, many people are larks, morning people like me, or owls, nighttime folks. Countless family conflicts occur for the good biological reason that people have different body clocks. Body clocks also change with age. Just as oldsters like to get to bed at rather early hours, adolescents, with their rapidly changing brains, prefer to go to bed considerably later. Such changes in body clocks can markedly distress families which don't take them into account, just as they do teachers, school administrators, and anyone who deals with kids.

To figure out what the different clocks are for members of your family, have everyone take the following test. (You've already seen part of it in determining your preferred waking and sleep times.)

Biological Time Test: Are You a Lark or an Owl?

1. Imagine you are greatly enjoying a vacation that lasts as long as you wish. You have no responsibilities, no worries, and more money than you'll ever need. You can do whatever you like.

   What time would you go to bed?

Between 8:00 and 9:00 p.m.	6 points
Between 9:00 and 10:00 p.m.	5
Between 10:00 and 11:00 p.m.	4
Between 11:00 p.m. and midnight	3
Between midnight and 1:00 a.m.	2
Between 1:00 and 2:30 a.m.	1
After 2:30 a.m.	0

2. You're still enjoying your very pleasant, unlimited vacation. Considering only your personal desires, when would you wake up?

Before 6:00 a.m.	6 points
Between 6:00 and 7:00 a.m.	5
Between 7:00 and 8:00 a.m.	4
Between 8:00 and 9:00 a.m.	3
Between 9:00 and 10:30 a.m.	2
Between 10:30 a.m. and noon	1
After noon	0

3. Though still enjoying your vacation, you're beginning
   to get a little stir-crazy. You think you want to start a
   volunteer job. It's work you've done before and that you
   really enjoy. You don't plan to overdo it. You will work
   only two hours at a time, continuing only if you find
   your tasks rewarding and entertaining. When would you
   pick your two-hour shift?

Between 5:00 and 7:00 a.m.	6 points
Between 7:00 and 9:00 a.m.	5
Between 9:00 a.m. and 1:00 p.m.	4
Between 1:00 and 7:00 p.m.	3
Between 7:00 and 11:00 p.m.	2
Between 11:00 p.m. and 1:00 a.m.	1
Between 1:00 and 5:00 a.m.	0

4. Your vacation is providing you relaxation, rest, and
   a profound sense of peace. Remembering the very
   different circumstances of your previous life, you are
   reminded of the times you felt free and at your best.
   At those times you would have described yourself as:

Definitely a morning person	6 points
Probably a morning person	4
In between a morning and a night person	2
Very much a night person	0

Add up your score and enter it here: ___

If you scored between 16 and 24, you are a lark. If you
scored between 0 and 8, consider yourself an owl.

If you scored between 8 and 16, consider yourself a
sparrow. You are in between, the silent majority.

Overlap Times

If all of your family members are sparrows, well and good. But what if there are two larks, a sparrow, and a brace of owls?

Welcome to overlap times.

Most sparrows find themselves sleepy in the midafternoon. They generally find themselves more alert and able to complete difficult mental, and some physical, tasks in the late morning and early evening.

Fortunately, the times when sparrows are alert and are best able to get things done are also times when both larks and owls are *relatively* alert. Though some larks will become quite sleepy around noon, they are generally up and about in the late morning. Most owls, even those who like to get up in the midmorning, are still able to talk and appear somewhat sociable by late morning.

Early evening is another good biological clock time for almost everyone, lark or owl. Early evening is when people generally experience their best moods; have the most accurate, powerful bodies (something to think about if you're trying to set a world sports record or even just a personal one); and feel the greatest degree of mental alertness.

For many families, both late morning, as it shades into early afternoon, and early evening become ideal overlap times. In particular, early evening may be a period when social connections may be easier to forge, when everyone in the family, unless bowed down by multiple jobs or shift work, has the chance to be at relative biological prime times for mutual engagement.

Of course, body clocks also affect performing all the many different types of active rest, physical, mental, social, and spiritual. They also provide biological and social underpinnings for using the different techniques of rest at the most effective times.

Some types of rest, like deep breathing, can be used anytime you are awake. In the list below, these are marked with (A). However, some of the many techniques of rest you have learned still have preferential times.

Here's a rundown:

Physical Rest Techniques

1. *Deep breathing* (A). Though deep breathing can be used at any time, it is often useful for calming yourself down from stress and preparing yourself to sleep at night. Work sometimes becomes more stressful in the late morning before lunchtime or when meeting deadlines. The time just before leaving work in the late afternoon is a very good time to do physical rest techniques like deep breathing, which will act as a bookend to the day, preparing you to go home.

2. *Mountain pose* (A). Mountain pose is surprisingly useful and effective at pretty much any biological time you want to use it. With practice, it can also be used effectively to help prepare you for sleep.

3. *Gravity pose* (A). Gravity pose is useful as part of a sleep ritual or to help set up a quick midday nap. It can also be used whenever stressed out or as preparation for periods of focused concentration right before difficult mental tasks.

4. *Brief naps.* Generally best in the early afternoon for larks, in the early to midafternoon for sparrows, and in the late afternoon for owls. Unless you are a shift worker, you *don't* want to nap in the evening, when, for most people, naps will interfere with nighttime sleep. UnNap naps, however, work well in the evening.

5. *Hot bath.* A great idea before nighttime sleep, hot baths can also be used to wind down in the evening or late afternoon after a really hard day at work or school or to relax in the afternoons of a calm weekend.

Mental Rest Techniques

1. *Self-hypnosis* (A). You can use self-hypnosis anytime. However, it's especially effective when you want to focus on a task or job. It can also help you fall asleep.

2. *Focusing the eye* (A). Focusing the eye is especially helpful when you are losing sharpness in the afternoon due to normal biological clock declines in alertness. However, it can also be used to simultaneously relax the body and wake up the mind at any time of the day.

3. *Walking to music* (A). Some kids will use walking to music when they wake in the morning, when it is generally hard for them to wake up. Many teenagers use it when hanging out with friends. Adults can walk to music when they're going to lunch or any time they need to obtain a better state of concentration and alertness. Office corridors are good places for walking to music.

4. *Ear popping* (A). Though useful anytime, especially in stressful situations, ear popping works well when you are making social transitions, such as when you are ending lunch or finishing the workday and preparing to see your family.

5. *Garden walks* (A). If possible, an early morning walk in a garden can wake up one's cold brain and lead to better weight control, just as an early evening walk can mentally and physically separate work and family life. Fortunately, garden walks have near endless uses, for both social and spiritual connections.

Social Rest Techniques

1. *Brief special connections* (A). You want the opportunity to use special connections any time of the day or night, as they are meant to be a part of handling emergency situations. Though

worthwhile if you have appropriate opportunities to practice them at work, special connections are most often scheduled for after work, in the evening, or during weekends, when more intimate family and friends are at home.

2. *Visiting a coworker or neighbor* (A). Though an anytime technique, especially for weekends, it's fun if possible to visit coworkers in the midafternoon, when biological alertness is lagging. Social connections at such times are very restful and can help you maintain greater mental alertness afterward.

3. *Quick connects* (A). Most quick connects will be made in the evening, though between both adolescents and loved ones they will be used throughout the day, particularly midafternoon—and, too often, throughout the night.

4. *Walking to lunch with a colleague, friend, or neighbor.* By definition walking to lunch is an afternoon experience. Walking to a meal in the evening has other advantages, as we will soon learn.

5. *Walking with a friend in a park* (A). Though potentially an anytime social rest technique, most social walks for working people tend to take place at lunchtime, in the evenings, or on weekend afternoons.

Spiritual Rest Techniques

1. *Praying for one minute* (A). Many like to pray for one minute as they wake up or right before they go to sleep. Others like to use it if they feel stressed or whenever they wish to center themselves and obtain a sense of peace.

2. *Moving through time* (A). Moving through time is an excellent technique when feeling overstressed at work or to regain focus at the end of the workday. Because of greater mental alertness, many also like to use it during the early evening.

3. *Moving through space* (A). Often used for stress control or in the early evening, moving through space is generally done during the same time niches as moving through time.

4. *Contemplating suchness* (A). An excellent technique in the late morning and early evening when one is naturally alert, contemplating suchness can also prove helpful as part of a sleep ritual.

5. *Separating subjective and objective* (A). Hard to do on awakening unless you quickly become alert, separating subjective and objective may work best at programmed times in the afternoon or evening, when interruptions may be less likely and one can more easily focus and concentrate.

There's much that can be done with all these rest techniques. However, rest at home allows for special social connections. Here are a few you can try.

DAY 24: REST AT HOME AS A FAMILY TECHNIQUE 1—THE MORNING MEETING

Every human is a unique experiment. That includes identical twins, who have different experiences beginning with their first few hours in the womb. Our environment produces differentiating changes even in those with exactly the same genes and homes.

As we are all unique, so are our waking and morning rituals. Children must go to school, adolescents must choose clothes, parents must figure out how to get everything done and eat breakfast.

Though eating breakfast as a family is an ideal, it is less commonly practiced now than it was in earlier times. Still, it pays to get the family together whenever possible, including mornings filled with rushing to school and work.

This can be accomplished through the morning meeting. The morning meeting is simply that—everyone gets together in the same place at the same time. You greet each other, look at each other. Then every person explains where they will be and what they will be doing that day.

The morning meeting is a good time to set up plans for quick connects (social rest technique 3). Phone calls, e-mails, or text messages at set hours let parents and children know where and what is going on, most important when everyone is flying off to do their own thing. Such brief social connections, especially when planned ahead of time, can calm parents and provide structure for children. Such social rest connections can also be done, if desired, very quickly.

If everyone's plans are regular and known to every other family member, the morning meeting becomes a chance for people to declare what they ideally hope to experience or learn that day. Children can mention what they would like to discover or simply whom they wish to see. Parents can humorously grouse and declare they hope to get through the workday unmarked and reasonably exuberant or describe how they will enjoy conversing with the friends, relatives, or colleagues they hope to meet.

Though many people protest they have no time for such things, morning meetings can be really brief. At the beginning, aim for short meetings—one to two minutes. If everyone has the opportunity to talk and can easily discuss things together, then take five or ten minutes. Good places for your morning meeting are the kitchen or dining room. These spaces perform better as meeting arenas than quick communicative gasps in front of the door leading from the house.

Morning meetings structure the day, provide comfort and stability, and increase family efficiency. They're worth a try.

DAY 25: REST AT HOME AS A FAMILY TECHNIQUE 2—MAKING THE EVENING MEAL TOGETHER

Though school, work, playmates, lessons, and the necessities of economics may prevent it, getting together as a family in the evening provides a worthwhile resting time.

Evening meals are fine overlap times, when owls, larks, and sparrows are usually all in relative sync with one another. People are alert, and moods are often the best they will be during the day.

Moreover, there's a lot to talk about. It helps to start by having everyone describe something that happened during the day, an experience they found meaningful. If you don't have such an experience to relate, that's okay. But learning something new, meeting a new friend, or conversations with outside relatives are generally topics everyone in the family can be interested in.

It's also entertaining and helpful to talk about the food on your table. Food is more than fuel; it's information. Food gives many, many different messages to your body, affecting mood, weight, and sleep (if you're interested, please see the section on food, activity, and rest—FAR—in the final chapter).

Food also can mean cooking a meal—together. Though lots of families like to eat out, it's often more fun, if more work, to cook and dine as a family. You can pick the menu together, explaining to the children where the food comes from and how it's grown. They can also look up each food's nutrients and what they provide to the body. As kids grow older, they can help more and more with the cooking.

Kids can also aid in washing up. It's easier doing what might otherwise appear drudgery when you're talking to people you love.

Cooking and eating a meal together provides more than social connections between family and friends. Dining is an innate pleasure, often mentally restful, and eating is one of the great joys of living. Dining together is an experience in which we can learn a lot, both about how we are what we eat, and how social eating can make life more enjoyable and meaningful.

DAY 26: REST AT HOME AS A FAMILY
TECHNIQUE 3—TAKING AN EVENING STROLL
TOGETHER

In Italy and Spain and other Latin countries, a social ritual is the family evening stroll. The family and the community come together at the same time, and in the same place.

People walk in the square or in parks. They greet their neighbors, say hello, and pass by, or they walk or stand together and chat. As they stroll and talk, they learn of political changes in the neighborhood— who has gotten well or ill, which folks are moving in or out, and how the economy has shifted.

Having a place to walk, especially a green space, is very important to health. Recent British studies looking at 360,000 people found health status improved, particularly for the poor, the more green space they had around them to use. Often the heath improvements were far greater than what is typically seen with major new health care reforms or after providing new health services.

Strolling as a family has other benefits: (1) exercise improves alertness and helps prevent obesity, hypertension, and atherosclerosis; (2) sunlight can improve mood; (3) you can meet neighbors and find out what is going on in the neighborhood, increasing healthy social connections; and (4) evening walks can promote better sleep.

Evening strolling also can make catching up with your friends and neighbors a very relaxing experience. Strolling acts for many as a form of mental rest and, when combined with social conversations, acts as a useful form of social rest.

Different environments also create different psychological states. Walking in greenery or a park often helps people feel more alert and alive and more attached to the natural world. Families can deepen their bonds when they walk together in a park.

Strolling outdoors as a family may be hard to do in the North during a snowy winter or in the South's steamy summer. However, acclimating to a climate also has its virtues.

The novelist William Faulkner complained about air-conditioning as "abolishing weather." Unless it's truly too cold, too wet, or too hot, walking outside as a family helps you control your response to weather rather than allowing weather to control you. Walking should not be a painful or difficult experience, and social networking may improve as you increase the size and nature of the realm in which you move.

Evening strolls have been used for millennia by different cultures precisely because they are so useful. At one stroke, they create conditions that can put together mental, social, and, when done in natural or historical settings, spiritual rest.

DAY 27: REST AT HOME ALONE— TAKING A WALK IN A PARK

Solitude need not be lonely. In the right dose it can be liberating.

Even members of the closest families like to be alone sometimes. Solitude lets you think for yourself, gather your thoughts and experiences, and figure out what you want to be and do.

Solitude can also help connect you with the wider universe, starting with the natural world. For many, walking in a park becomes a spiritual experience, itself a form of spiritual rest.

Different biological times change how you feel and what effects walking in a park will have on you. Walking in the park in the afternoon during the workday can provide great mental rest. If your employer is interested, you can tell her it's been shown that afternoon exercise, like walking in a park, can markedly increase work productivity, perhaps by overcoming the normal biological clock torpor that occurs in the afternoon. The physical and mood benefits of movement and light also apply.

An evening walk in the park is different. As the day wears down, light shifts and moves and sometimes becomes furtive. Plants change their appearance, as trees create silhouettes in evening light whose complex, branching patterns may visually enchant you with a gorgeous form of line and rhythm.

Smells also change come evening, as do colors. The eye shifts during the night from emphasizing retinal cones to relying on black-and-white-sensing retinal rods, changing the interrelationships of reflected light in entertaining ways.

Evening walks in a park also help to separate work from what comes afterward. The day gets unconsciously summarized and remembered as you walk from plant to plant, an act that itself may prove meditatively restful.

Walking in a park can provide mental rest and spiritual rest. And if you meet a friend, it can provide a social connection for you in a special place. Sometimes the connections you forge with strangers who are interested in the same natural landmarks as you can last much longer than a single, solitary walk.

Summary

Rest at home allows for an extraordinary freedom to combine social, mental, physical, and spiritual rest. Family members and friends can work together to forge new bonds in environments that are safe, stable, and enduring.

In this chapter you've learned about how needs for rest change with aging and shifting biological clocks. You've learned different techniques of how to

have a morning meeting as a family

make an evening meal together

take an evening stroll together

take a solitary walk in a park

Overlap times allow families that otherwise seem separated to come together as a unit, conversing with, cajoling, and learning from each other. Evening meals and strolls can provide special forms of communication that can open up further connections. A simple walk in the park can renew mind and body, amusing and entertaining us as we put together our thoughts and experiences in fundamentally different ways.

The different types of rest are indeed skills that improve with time and age. They also improve as we do them with others, learning from them to create new ways to perform the simplest acts of living. Social rest connections forged at home can create a sense of well-being and peace that can keep the worries of the wider world effectively at bay while connecting us more deeply to the ones we love.

chapter 8

REST AT WORK

"How can I rest? There's no time. Rest at work? You must be crazy."

Not at all. Human beings are not machines. Our bodies are built to do things well at some times and to rest at other times.

Rest at work has many benefits. You can revive yourself in the middle of a tough day, relax in the middle of great tension, and do simple, practical things that not only will help you feel better but will also make you more productive.

Still, getting rest on the job, even though useful, is sometimes hard to do. There's simply too much to do.

Unless you get enough rest.

Many studies now argue that social rest activities—simple things like taking a walk at lunchtime with a friend—markedly improve work productivity in the biological clock slow zone of the early to midafternoon. Social rest also connects workers together and, beyond providing common purpose, can reknit together a malfunctioning corporation or public institution. Different kinds of social rest can also make for more creative ideas within companies, along with more creative workers.

However, time is normally always an issue at work. First, people feel they don't have enough of it to get done what they need to do. Second, they don't always realize just how much biological clocks affect performance, particularly in stressful environments. Third, finding a place to rest at work may demand quite a bit of ingenuity.

Let's go through these three different work issues, starting with the problem of time pressure. The failure to allocate time for rest, restoration, and renewal is a major one. That is one reason why many of the physical, social, mental, and spiritual techniques described in this book have been designed to be done expeditiously. With practice, most of them can be done within a minute.

Let's go down the different rest techniques and see which can be accomplished within sixty seconds.

Quick Rest Techniques

Physical Rest

    deep breathing

    mountain pose

    gravity pose

Mental Rest

    self-hypnosis (with practice)

    focusing the eye

    walking to music

    ear popping

Social Rest

    brief special connections

    visiting a coworker or neighbor (better if done for two or three minutes, but this can in a pinch be done in one)

    quick connects

Spiritual Rest

    praying for one minute

    moving through time

moving through space

contemplating suchness

separating subjective and objective

With practice, rest techniques can be done quickly and effectively. Some people complain that there are, if anything, too many different rest techniques to use and combine, an embarrassment of riches we will address further in the final chapter, "Tuning Your Life." What's important to know is that there are many different rest techniques you can use in a short space of time; that most can be used almost anywhere; and that as you use different techniques, you will quickly discover the ones you like best and when during the day they work most powerfully for you. Combining rest techniques is a major way to increase flow activities in your life and to make both your work and leisure days rhythmic and musical.

Power-Ups

Sometimes you feel overwhelmed. There's too much to do and not enough time to do it. You need a quick reset or a hit of energy to get you through the project. If your need is great and time short, these Power-Ups can provide the zest you need. Try these rest techniques or combinations.

1. *Ear popping.* If your coworkers make steam come out of your ears, find a few seconds to pop your ears. If possible, close your eyes for ten seconds as you listen to the silence your fingers blissfully induce. Then pop your ears loudly, and look anew at the world around you.

Concentrate first on color alone. Name the colors you see. Then look at the shapes of objects and what they are. Next, feel the size of the room you're in. Then listen, trying to identify each of the individual sounds around you. In a few seconds you should have reset your perceptual system, getting you ready to move forward.

2. *Deep breathing in mountain pose.* Stand straight and tall. Align

your ankles, knees, hips, and shoulders. Breathe in to the count of four, out to the count of eight, hearing and visualizing the moving air you breathe. In a few seconds you may feel yourself as quiet as a forest and as immovable as a mountain, ready to do what needs to be done.

3. *Walking to music.* Pick a fast tune you love, and walk to it. Feel your legs and arms respond to the fast tempo, and accentuate the downbeat as your feet hit the ground. Move like you're dancing, but dancing purposefully, excited to get to your next project.

4. *Making a brief special connection.* You've got something overwhelming to do and very little time in which to do it. Despite the time pressure, call someone you trust for a brief special connection. This person already knows you may call in times of travail. If you have the chance, ask his or her advice. If you don't, just ask for their support, and know there will be others there to help you.

5. *Self-hypnosis.* Your problem appears large, perhaps too large to finish on deadline. Quickly put yourself in a self-hypnotic state of relaxed concentration. Imagine what the job entails, and then visualize yourself doing it quickly and efficiently. If you are too anxious to do self-hypnosis, imagine you are your identical twin living a thousand miles away. How would you complete the task? When you open your eyes a minute later, you should know what to do.

Where to Rest on the Job

Finding a good place to rest can be a little difficult in many workplaces. If you're a schoolteacher with thirty-three first-graders, any lapse of attention may provoke unknown consequences. People working in call centers who are monitored second by second often can't imagine that there's any place they could possibly rest.

Remember this: thought is action. What you do with your brain changes how it works—chemically, physically, even anatomically. To rest in difficult work circumstances, you have to use your head.

Consider the different forms of mental rest. Self-hypnosis can be induced quickly and effectively within a minute (as we'll show you below, in big screen technique). Focusing the eye can be done in twenty to thirty seconds while seated at any desk or chair. Walking to music can also be done in spurts of twenty to thirty seconds, accomplished while walking over to your supervisor's office. Remember, ear popping can and should be done anywhere.

Think a moment about the different techniques of spiritual rest. Moving through space and moving through time are techniques that require mental concentration but can be done, with practice, in most any workplace. Both are techniques you can do anytime and anyplace, as is contemplating suchness (spiritual rest technique 4).

Even people who feel attached to their workspaces like glued-in-place mollusks have to get away sometime. Everyone has to go to the bathroom. That brief stroll can be used to walk to music and can also include an energizing, brief visit to a coworker.

Standing waiting for the elevator? A great time for mountain pose. Answering a phone call from an impossible client? An opportunity to use deep breathing.

If you are lucky to have your own space at work, you can start to do things that may appear at first unthinkable to you—like taking a nap. Finding places for other rest techniques, like gravity pose or a rapid afternoon garden walk, may involve no more than a thirty-second conversation with a boss or just a somewhat longer bathroom break. Such chances to rest can also move you a bit out of your comfort zone, an opportunity that can prove surprisingly helpful in meeting others at your worksite.

So, yes, there are times to rest at work. Even for monitored phone workers, there are places to rest. But when are the best times to rest?

Body Clocks at Work

Your body never stops working. As you now know, the longest rest experience, sleep, is remarkably productive in creating new memories, aiding learning, and rebuilding your cells throughout the body and brain.

Like sleep, body clocks have powerful effects on our lives, whether the clocks are very long or very short in duration. People often feel more energized and physically active during the spring than in winter, as the seasonal effects of light change mood and with it subjective mental and physical capacity. Women are very aware that monthly menstruation changes how their bodies and minds feel.

Yet perhaps the most important body clocks for the workplace are the twenty-four-hour ones, which dramatically influence human performance. Body core temperature can be used as a guide, with an approximate lag time of two hours. Body core temperature up—greater alertness. Body core temperature down or flat—less alertness and sharpness. This has several implications for work:

1. For most people, alertness peaks in the late morning and early evening.

2. A large part of the normal workday, from the early to midafternoon, is a period of relative slowness and lack of alertness.

3. Sleep-deprived Americans should expect to be slow and far from their peak mood and performance at the start of the workday.

In other words, from the point of view of body clock performance, much of the standard workday occurs when people are *not* sharp.

That's where the power of rest arrives to help you. Rest techniques can, with proper timing and sequencing, reverse crankiness, fatigue, and dullness and liven up the workday. Let's see how different kinds of rest can ride to the rescue at the different hours of work.

Timing Rest Right During the Ten Hours of Daytime Work

Except for people who do shift work (an issue addressed more thoroughly in my book *The Body Clock Advantage*), most Americans do much of their work between 8:00 a.m. and 6:00 p.m. Each hour has its own different needs and requirements. That fact lets you personally tailor your rest techniques to your job, hour by hour.

Let's take a closer look at each of the hours of the workday.

8:00 a.m.

Usually the start of the workday is hectic and a bit crazy. There's the requirement to wake up your brain from a probably less-than-full night of sleep, check out the problems and put out the fires left over from the previous day, plus the need to let other people know you're present, conscious, and ready to do what must be done.

You may already have enjoyed the advantage of the morning meeting (day 24's rest at home technique), which provided you with structure and helped you prepare for what you will ultimately do during your workday. But now that you've arrived at your workplace—by foot, bicycle, car, train, or bus—you want to quickly move on to action.

Americans generally arrive at work in cars. As you climb out of the car, look on the walk into work as an opportunity to rest.

You can walk to music (mental rest technique 3). If you want, use an iPod or equivalent, but you can also listen to the music in your head, which may already be playing there, even if you are sometimes barely conscious of it. As you walk into your job, pick a tune that's quick, catchy, and strong or just simply happy. You are, after all, on the rising part of your body core temperature curve, which is making you more alert and alive.

As you walk to that rhythm, try to let the song, melody, or riff move through your whole body. If it's a march, march to the beat. If it's a pleasant, upbeat love song, try to sense what deep social connectedness feels like in your chest and arms.

If you can, put a little dance rhythm into your step, even if the workplace you're entering feels unappealing that morning. Remember that with practice you can start to control your consciousness more fully, which can provide your body and mind much joy.

As you wait for an elevator or find yourself caught in line, stand in mountain pose (physical rest technique 2). If you see a colleague who interests you and you don't need to rush to your office, spend a few moments visiting with this person (social rest technique 2). Ask him how he and his family are doing. Or look at her face, the way she holds herself; does she look tired? Perhaps a joke will come to mind, something that can immediately connect you both.

As you arrive at your desk or workplace, it's time to put a couple of rest techniques together to start the workday.

DAY 28: REST AT WORK TECHNIQUE 1— VISUALIZING THE WORKDAY

You can visualize the workday while sitting or standing, though most people will perform this technique at a desk.

Begin with some simple deep breathing (physical rest technique 1). Open your lips. Breathe in to the count of four, out to the count of eight.

Visualize the air moving down your pharynx, into the trachea, quickly following its intricately branching bronchioles. Feel your rib cage move up and out. Let your mind visualize the air as it flows into your bright, luminous alveoli, enriching your blood with the oxygen that gives you life.

After two to three breaths, visualize in front of you a single piece of paper (you can also pick up a real one, if you like). Draw a vertical line through the middle.

On the left side, visualize your priorities for this workday. To start, just put down three. The first priority is important. Visualize the task, and try to see clearly in your mind what it demands.

Today may be the day when you plan to finish a project you've spent weeks working on. Or you may want to set up a meeting with your boss, who's thinking of expanding your division and will hopefully choose you to run it. Perhaps you've gotten a bad cold and intend to just get through the day alive and unexhausted and without infecting anyone else.

With your first priority clearly in mind, turn to the right side of the paper. That's your day, marked down in hours. Look at 8:00 a.m., 9:00 a.m., 10:00 a.m. For now, think of your time in hour-long blocks.

Think about how you can put your first priority into effect during each time slot. In that first hour you may have no choice but to put out fires and finish regular, essential tasks. But by 9:00 a.m., you should be set to work on your first priority. Try to get to it as quickly as you can.

If you can work on your first priority for much of the first part of the day, you may then have a chance of gaining a sense of real accomplishment.

Next, look at the rest of your day, hour by hour. See what your schedule is. Check the times when you can work on your second and third priorities.

Perhaps you won't be able to get to your first priority at all. If that is the case, it's important to know it right out the starting gate of your day.

Now, as you run through the hours of your workday in your mind, think of which rest techniques you'd like to try out, and when. Try to insert specific places for them in the schedule. Remember that with the right rest, you'll work more productively.

Finish visualizing the workday by repeating to yourself three things: (1) what your first priority is; (2) when you think you will have the most, and most effective, time to work on that priority; (3) the times and forms of rest you have picked out for the day.

Now you have visualized what you really want to do during the day and when you want to do it. We now move to the next hour of the day.

9:00 a.m.

Unless 9:00 a.m. is the start of your workday, by this hour most folks are fully engaged in their jobs.

For some, putting out fires is proving a bit rough. Ear popping (mental rest technique 4) will always be available. All you need to do is to put your fingers in your ears for five or ten seconds. If work conditions allow, you can also close your eyes.

When you open your eyes, look around. Look at the colors, the light, the shape of the furniture, then listen to the noises around you. Pay attention to all your perceptions, noticing how you see the world.

Now that you've reset your perceptions, consider the tasks ahead of you. Is there still too much to do? Can you cut the different tasks into doable pieces? Are you bothered that your boss seems really tetchy today? In a few seconds you should feel more grounded and ready to get back to the different tasks facing you.

Perhaps your day is instead going splendidly, and you feel pleasantly engaged and physically well. You're getting done all that you need to do.

It's still worthwhile to set a rhythm for the day and take a quick rest. Breathe deeply for a few moments, or focus your eye on something lovely in the environment (mental rest technique 2). Recognize that shifting gears, moving through the day with variety, can stimulate your capacities and stabilize a good, optimistic mood.

10:00 a.m.

For most of us except extreme owls, by 10:00 a.m. our body clocks are moving in a positive direction. There's a lot to do, and we're doing it.

Yet you don't want to stay too long in exactly the same place. Your body needs to rest, but it also needs some physical activity.

If your job keeps you in one spot for most of the day, it's a good idea to get up and move. This may be a fine time to walk over to a coworker and ask her about a project you're both working on. If that is not possible, you can consider a quick walking trip around or through the office. If you need an excuse, say that you need to go to the bathroom. Walking there quite literally transforms it into a rest room.

As you walk, change your musical pace to a different, enlivening tune. Greet people as you go. As you move through your office or workplace, imagine all the objects there—the concrete beneath your steps, the iron rebars in the walls, the infinitely branched particles of flooring, the many different historical forms used in the furniture designs, the sharp and shallow changes of moving light. It's a small reminder of suchness, the infinite variety of the world we can learn to contemplate in a moment. It's good to periodically rest your mind when at work with these brief, restful reminders of the great world beyond.

11:00 a.m.

Now you should feel a bit revved up. Peak morning alertness may well be upon you.

This is a good time to do big screen technique.

DAY 29: REST AT WORK TECHNIQUES 2 AND 3—BIG SCREEN AND TAKING A NAP

Traditionally 10:30 or 11:00 a.m. was the time when workers took coffee or tea breaks. Coffee breaks may be lost to the sad idea that humans are machines, but 11:00 a.m. remains a great time to reevaluate your day and reinforce your sense of purpose.

You start big screen technique using self-hypnosis (mental rest technique 1). First find yourself a comfortable place to sit. Sit straight.

Now put your left hand (right hand if you're left-handed) over both your eyes.

Raise your eyeballs, staring straight at the ceiling as you close your eyelids. Take a very deep breath. Hold it for a few seconds. Though your eyes are closed, you're still "looking" straight up at the ceiling.

Now visualize in your mind where you are sitting right now.

See the room, the walls and floors, the slight disarray on your desk and the god-awful mess on your coworker's, the crackling, yellowish fluorescent lights—all of it as if you are watching everything on a giant movie screen. Now, see yourself on that screen.

Breathe slowly and deeply as you calmly observe yourself on the screen. Have you managed to get your first work priority all done? Have you managed to work on it at all?

Consider what you have accomplished and what you really want to do the rest of the day. As you visualize the rest of the day, see yourself getting your tasks done. See the steps you'll need to take, the potential sequence of actions that will make everything flow more smoothly.

Your eyes are still pointing up, your hand is in front of your face, but you're thinking about and *seeing* what you want to do. You're also considering where you want to be as the day winds down. As your breathing slows think of how you will feel when you can accomplish your goals, the inner sense of satisfaction that will create.

To finish big screen technique, lower your eyes. Take one last deep breath. Open your eyes, and take your hand away from your face.

Reality may not always go as you would like, but you have now seen a way forward. Big screen allows you to evaluate your day, recognize its challenges, and define the skills you'll need to accomplish them—all while you're sitting in your chair.

Big screen technique has lots of advantages for almost any workplace. First, it can be used just about anywhere. That includes boring business meetings when you're watching several people produce nothing more effective than verbiage. At these times, use big screen to understand the politics of your job, what people want, and what techniques they are using to get what they want. You can also do a more individually centered big screen while on a bus or train or seated as a passenger in a car, presuming your grandmother who needs cataract surgery is not driving. You can do big screen wherever and whenever you feel safe, which hopefully means frequently. Like other rest techniques, it gets better and better with practice.

Other techniques can also be used to reassert purpose and focus as your body alertness reaches its first peak during the late morning. Techniques of spiritual rest, like moving through time and space, allow you to quickly calm, and they provide much perspective on where you are and where you are going. Late morning is also a good time for making a brief special connection (social rest technique 1), a fast call or e-mail to someone you trust, letting that person know you're there and that you care.

Noon

Noon is the traditional lunchtime hour. It's also a sanctioned time for social rest.

Techniques like walking to lunch or walking in a park with a colleague provide more than social connection. Many studies now suggest that walking or moving in sunlight improves productivity for the biologically more difficult early and midafternoon and that walking through greenery has special direct and indirect health and alertness benefits.

It's hard to walk in a park in cold wintry weather. Yet wintry landscapes have their own beauty, even when the land is not covered with fresh snow. People feel better when connecting with nature, sensing they are part of something larger than themselves while moving through a living environment that in many cultures represents life itself.

Even if you can't walk outside, the noon hour is an excellent time for mental and spiritual rest. If you don't have the chance to sit and talk with a colleague as you eat lunch, you can use focusing the eye to calm and clarify thought. If stuck at your desk, you may find it helpful to have some piece of nature on which to look and meditate. If you have a proverbial black thumb and kill cacti and jade plants within days, put some nonliving natural object nearby. It can be a crystal or an interestingly shaped rock you found on a favorite trip. Concentrate on it for a brief while, trying to see nothing else. There are many different ways to rest during the lunch hour.

1:00 p.m.

If you've rested well at lunchtime, had bouts of social, mental, and perhaps spiritual rest, then the afternoon should hold considerable promise of productivity. Yet for many people, except owls, who are just revving up, the early afternoon is a period of relative slowness.

If you can control your sequencing of work tasks, early to mid-afternoon is a fine time to do those things that demand less intellectual and mental concentration. Unfortunately, that's often a pipe dream. Instead, you may want to use specific rest techniques to stay focused, like self-hypnosis or quick social connects.

If you're feeling a bit tired and sluggish, take a minute to stand up and do mountain pose. Align your ankles, knees, hips, and shoulders along the same imaginary line, breathing in deeply and slowly. With successive practice, mountain pose will often help you stay alert. (If you know other, more complicated yoga poses, you can take a few moments to do them too.)

2:00 p.m.

Two to three o'clock is the witching hour for a large part of the population. Many people feel sluggish, tired, and sleepy.

It's a fine time for physical rest techniques.

Take a Nap

From chapter 3, you now know how to take an effective, short nap. Many people think taking a nap at work is impossible if not immoral. I disagree.

I first started taking afternoon naps between 2:00 and 3:00 p.m. when I was teaching at Brown University. My office was pleasant and comfortable and had a door I could close. It had no couch or obvious place to sleep, but it did have a carpet. The first naps I took were after late work nights, when my sleep had been truncated. I lay down on the floor and wadded up a jacket as a pillow.

I rarely used naps during those days. However, I eventually made short naps easier by adding a pillow for my head and a simple eye mask for my eyes—a folded washcloth.

If you have control over your workspace, naps can be easy to take. A yoga mat can be used as a futon; it's both cheap and versatile. You can keep a pillow on a chair, and use that pillow for your head, or if you prefer an air pillow can be blown up for the occasion. A simple night mask over the eyes can effectively block out even a bright sun's glare.

There are a few useful tricks to obtaining successful naps. First, if you are at work, make them short. Long naps—longer than fifteen minutes to a half hour—will tend to pull you into deeper stages of sleep.

You don't want that. Deeper sleep means sleep inertia, which will make you feel momentarily sluggish and dull when you wake.

To control the length of your nap, use a kitchen timer or a watch alarm.

Make your positioning comfortable, and try to make your space as quiet as you can. You'll probably have to turn off the tones on your computer and cell phones. When you are sure you won't have any calls, you can then use your cell phone or computer as your timer.

If you have lots of people reporting to you, see if they can leave you undisturbed for a short while.

No-Nap Environments

Though several studies show naps improve work performance, many bosses don't believe that to be true. You'll just have to prove it to them.

Let your employer know you won't take naps unless your work demonstrably improves by napping. If your work output is easily quantified, show your boss the difference by tracking your output one week without naps and one week with naps. It is usually possible to demonstrate your productivity edging up when you have the time to rest.

If you don't control any office space, you'll need to find a place to nap. Often workplaces have conference rooms that are largely unoccupied, or occasionally empty offices. That's another reason to visit coworkers during the day, as they can often help you out, including letting you in on good places to rest.

If you are like many American workers, you're not working strict eight-to-five or nine-to-five hours. If you are working night and day, maintaining full readiness at work may sometimes require you to take a nap.

You always, however, have the option of trying an UnNap nap, as described in chapter 4 on mental rest. UnNap naps have many advantages. They can be done while seated at your chair and often require only a minute or two to let you reach a state of real mental rest.

If instead you really need to quickly wake yourself up, physical and social rest techniques that involve walking, especially in sunshine, usually are preferable. In the early to midafternoon, Power-Ups can be useful for that purpose.

3:00 p.m.

Night-shift workers tend to hit the wall at 4:00 a.m. Day-shift workers often find 3:00 p.m. a difficult time.

Alertness is generally improved by socializing. As the afternoon hits midstride, think about visiting work colleagues, or briefly walk to music. Midafternoon is also a time to reintroduce a potent technique of the workspace prior to the Internet era—the coffee break.

DAY 30: REST AT WORK TECHNIQUE 4— COFFEE OR TEA BREAK

An excellent time to take a work break is at 3:00 or 3:15 p.m. If done right, the break will prove restful and help you work more productively. Coffee or tea may be served during your work break but are certainly not essential.

When I was young, I studied in England at the University of Sussex and later at Cambridge. At Sussex I was enrolled in the School of Molecular Sciences and was deeply impressed by the productivity of British labs.

The English scientists worked differently. The American labs where I had labored seemed to operate on a principle of brute force— if you could do experiments for twenty hours a day, it was twice as good as working only ten hours a day.

The British approach was nearly the opposite. Long hours were almost frowned upon, which did not prevent people from coming in during the weekend to "just see how things were going." At least from my lowly perch it appeared that there was less tension in the English labs than the American ones.

Dutifully at 11:00 a.m. and around 3:30 p.m., those who could get away would temporarily quit their tasks and have a cuppa tea. Graduate students and postdocs would gather around their lab chiefs, imbibe caffeine, and talk.

Sometimes the conversations were pleasant small talk. Much of the time, however, people discussed their work—what they had done, what they hadn't, and what seemed to be effective.

People automatically cooperated at such times, answering each other's questions in a relaxed manner very different from the American weekly lab meetings I had known. They also took the trouble to gather together people from separate labs.

The Molecular Sciences Department at Sussex was distinguished. Several of the faculty were fellows of the Royal Society, including one rarely seen Nobel Prize winner. They seemed quite at ease going over to each other's facilities and talking about their challenges. Though sometimes their research areas were quite dissimilar, people took the time to air out their ideas with colleagues.

Later, Silicon Valley would follow similar procedures, which helped make many companies remarkably productive. One of England's greatest poets, John Donne, was correct: no man is an island.

Take your coffee or tea break alone if you must, but try to visit with work colleagues if you can. They can be your immediate coworkers, but that is certainly not required. Often it's fun to take a break for five to fifteen minutes with someone from your workplace who is not your usual coworker, someone you might otherwise hardly see.

You then have a chance to socially connect, allowing you to achieve some social rest. Finding out what's happening elsewhere in your workplace can improve your morale and that of the people you're speaking with, as information gets passed from person to person the traditional way, face-to-face. You will also break from work at a biological clock time when overall productivity tends to be low.

Rest is renewal. If you don't have the chance to socially connect with coworkers, you can still brew that nice mug of green tea. Savor it. Then you can concentrate for a moment on focusing the eye, helping you to quickly put your mind at rest.

4:00 p.m.

By late afternoon, most people's innate alertness is rising. At such times, it's fun to use quick social connects. To make life rhythmic, it's also fun to communicate, if appropriate, with people outside of work.

If you have the opportunity, this hour can also be used for blue-sky kind of thinking, where you open up yourself to imagine thoroughly innovative plans or projects. A quick walk, especially in a garden, can let you use all the experiences of the day to come up with new ways to tackle your job challenges.

5:00 p.m.

Before you leave work, you want to take stock.

Though you may be excited to leave the workplace and see loved ones, take a short time if you can to do deep breathing (physical rest technique 1) followed by visualization.

Once you feel relaxed and concentrated, think of what tasks you've managed to finish during the day. Did you get to your first priority and advance it? If not, try to think of ways to do it your next workday.

Next, see what you learned on this day. It may be a new technique you picked up from a colleague or that your boss is a real bear if you try to talk to her just before she runs to lunch. Consider what really stands out from all the other actions of the day.

Last, visualize the rest of your daylight hours. Imagine the possibilities and the pleasures you look forward to. Soon you should be able to put them into action.

The finish of the workday is also a good time to use spiritual rest techniques. You can move through time and space or take a few moments to focus on separating subjective and objective. It's often good to sense your connection to the greater world at the end of a workday, providing you another chance to feel a sense of accomplishment about whatever you've done. It's fun to use spiritual rest techniques at the end of work much like a coda that comes at the end of a piece of music. The storm, stress, and achievements of work may now lead to something quieter and more peaceful as you become more aware of where you are and where you've been.

Summary

Rest at work is both possible and necessary. Though finding time and place may take a little ingenuity, rest at work restores you and provides a rhythmic pattern to the day that increases productivity and your personal sense of achievement and enjoyment.

In this chapter you've learned four separate work rest techniques:

1. *Visualize your workday* before you begin so that you can define your first priority and a way to achieve it.

2. *Big screen technique* will give you the chance to obtain calm and perspective in a matter of seconds while also giving you a fine view of your place in office politics.

3. A *short nap* may revive you out of the biological dead zone of the early to midafternoon.

4. A *quick coffee or tea break* may not require either beverage but can provide the planning, cooperation, and social connectivity that can help you and your organization become more productive and socially connected.

At work, at home, or at play, rest is restoration. The many different rest techniques you've learned can be matched and combined to fit into almost any workday or environment, providing a sequence of action and rest that can minimize stress and maximize pleasure.

part 2

PUTTING IT TOGETHER— MAKING LIFE MUSICAL

chapter 9

SEQUENCING

How to Deal with Multitasking and Boredom Through Flow

Life is short. Why do one thing when you can do two? Or three? Or five?

Multitasking lets us do so many things at the same time. When we multitask there's novelty and buzz as we juggle more and more balls in the air. Multitasking slams together ordinary tasks like an amusement park thrill ride, daring us to see how much we can do and how far we can go. No wonder so many people complain about multitasking and then do it so frequently. Multitasking can be fun.

It's just that the human brain is not very good at it. Human design shows that no matter what we think we're doing, the human brain really does things pretty much one at a time.

Luckily there are options to multitasking that are both more effective and highly enjoyable. Mihaly Csikszentmihalyi's concept of flow allows many of the most boring tasks to be transformed into interesting activities and provides an easy guide for using concentration to overcome distraction. The payoff is we become more amused and more productive at the same time. Next, if we marry flow with proper sequencing, lots of things that seem impossible become feasible. Sometimes, with practice, they even become easy.

People multitask for a reason—they enjoy it, or, in busy work and family environments, they feel they have no other choice. And if you don't have enough time to do what you absolutely must do, when is there any time to rest?

The answer lies in understanding how the human body really works and allowing its innate rhythms, cycles, and inner music to work for you, not against you. We need to be active and to rest, renew, and restore, and we need to know how and when to do both. That gives us the chance to get a lot done and also have a really good time.

Multitasking and the Pleasures of Sensory Overload

The one-ring circus led to the three-ring circus and then to the five-ring circus. If you play a video game you may be attacking and fighting off alien monsters as you plan your strategic thrust on the hidden imperial base while simultaneously trying to save your ambushed colleagues until the missile from nowhere blows you up. Attending a rave, you can detonate your brain on ecstasy, overloading your neural circuits with too much information as you wildly dance to techno whose sounds vibrate through your eardrums as the beat penetrates your bones.

Lots of us enjoy sensory overload. It's a thrill when everything happens at once. Yet there are other reasons people like to multitask.

Recently I visited the house of a twenty-something painter friend to work on a cross-cultural arts project between China and the United States. He was not in. His roommates were—two smart and funny women recently out of college. They invited me to wander in to their charmingly ramshackle house and sit down and converse. Sort of.

I rolled into an old chair while the two sat across from me on the northern and southern ends of a yellow cloth sofa. Leaning forward, they glanced at me above matching laptops, which they continued to observe and type on as we talked.

Because I come from a different generation, my first internal reaction was a vague feeling they were being rude, but I checked myself and soon realized that I was wrong. My two acquaintances were doing with me exactly what they did with their other friends—surfing the Net, typing e-mails, getting in an instant message or two as we talked. On occasion they did a search on some item I mentioned and then returned to the conversation.

I asked them why they were writing, messaging, watching videos, and reading at the same time they were conversing. They were surprised by the question. They explained that what they were doing that afternoon was normal for them, entirely routine, and kind of enjoyable. Then one said she enjoyed performing several tasks at once because "it's not boring."

In my clinical experience, boredom scares people. They begin to feel separate, untethered, lonely. When you're bored, all the internal demons lying in wait feel the itch to escape. Suddenly out of nowhere fly up the pains of the past, unquestioned fears of unmet expectations, anxiety about the future. People ruminate and worry as if they are standing in some endless airport security line, not knowing if they will be asked to walk through an unmarked door.

In the United States we've been taught to work hard to avoid boredom. Boredom is the enemy. It's no surprise many kids write hundreds of text messages every day.

My parents' generation had books, radio, eventually television. In those days long past, people generally knew their neighbors, often visited neighborhood friends, and strove to entertain themselves.

Today people can be vicariously entertained anywhere they choose. Few spots on the globe lack multiple options for amusement. All you need are batteries and something with a silicon chip, whether it's a cell phone, a notebook, a handheld, or a sleek electronic reader. There are too many paths to immediate entertainment to take the time and imagine what has been lost.

Distraction Versus Concentration:
The Problem of Multitasking

Multitasking is not new. Humans have been doing it for thousands of years. We have even studied multitasking for well over a century.

What do those studies show about how multitasking operates in the brain? Harold Pashler of the University of California–San Diego wrote a fine paper in 1994 going over the more than a hundred years of psychological studies of multitasking. The issue, in psychology research terms, was called dual-task interference.

We do two things at once so often that we hardly notice. We walk down the street and chew gum. We stroll through the neighborhood conversing with a friend. If this is multitasking, it can't be that hard.

You would expect that multiple actions are so easy and ordinary that it would prove difficult to demonstrate the effects of dual-task interference. What Pashler and lots of others have known for decades is just the opposite: doing two very simple tasks at the same time really screws up human performance.

Even something as simple as pressing a button when you hear a buzzer—what's called reaction time—is markedly changed by dual-task interference.

The classic experiments were performed by Telford in the early 1930s. Using a buzzer, people would respond to two stimuli, one directly after the other. Then he looked at their response times.

What happens is that as the intervals between the two sounds decreases, people's response to the second sound *slows*. Put two things together, and the closer they occur to each other, the worse the response. Telford and colleagues named this phenomenon the psychological refractory period.

We bump up against the psychological refractory period all day, every day. The question is, what causes it? Through many years of experiments, people figured out that some kind of bottleneck was occurring in the brain. Something slowed the whole process down when you moved from one task to another.

Experiments done by David Meyer in 2001 provided some insight into why. His research group proposed that two things need to take place when you go from performing even the simplest task to performing a second. First, parts of the brain have to "goal shift," meaning you have to turn on *and off* different parts of the brain so you can (1) pay attention to what's going on and (2) go out and perform the second task.

Next, the brain has to do something called "rule activation." This is when your brain, through calculating probabilities, orders and begins its next sequence of moves. If I'm going to chew gum *and* walk simultaneously, a whole set of rules must be accessed to determine what parts of the cortex must turn on to keep the muscles of mastication operating at the same time as those of locomotion. If I then try to walk, chew gum, and also *talk*, a whole new set of goal shifts and rules is suddenly required.

What feels so easy and automatic to us is to the brain anything but simple. Much of the business of multitasking is thought to take place in the prefrontal cortex, which in human beings takes up about twice as much cortical space as it does in chimpanzees. The prefrontal cortex is said to act to aid executive function by the brain, which in turn is considered the executive organ of the body.

Of course it's vastly more complicated than that. The prefrontal cortex has to communicate continually with the frontal cortex, where many muscle actions are coordinated with our conscious, long-term goals. In this case, the reason we're walking is that we want to get to the restaurant to eat pizza. Other "quiet" parts of the brain like the cerebellum also play their complicated roles. Many, many brain systems have to work in concert if you're going to do something really complex like walk to the mall with a friend.

And no matter how many tasks are required, they will mostly be done one by one, in sequence. Work by Earl Miller at MIT demonstrates that even multitasking people are only focusing on one or at most two things at a time—period.

So you can imagine the results when people try to combine actions that are not as routine and automatic as walking, a critical physical process that has specially evolved for millions of years. When David Meyer had his students switch from math tasks to writing tasks, their overall performance declined. A lot. The more complex the performance required, the worse the results.

Things worsen further when you look at actual workplace studies. Research by Gloria Mark at the University of California–Irvine shows that present-day workers are generally staying on task for only eleven or twelve minutes before they are interrupted or interrupt themselves. Getting back to doing the original job usually takes more than twice the eleven or twelve minutes they spend on a specific task.

Even if we enjoy doing it, multitasking means less work gets done. These days multitasking is sometimes fatal, especially when we're engaged in mechanical transport. I think of the Los Angeles train conductor who died in a massive wreck in 2008, text-messaging while he drove, or the fourfold increase in accidents when people drive while talking on their cell phone. Multitasking fouls things up big-time. Texting and driving is a special disaster.

Even more odd, recent research shows that multitaskers are actually worse at multitasking than people who don't generally engage in it. Clifford Nass and researchers at Stanford reported in 2009 that not only were high multitaskers lousy at performing multiple tasks done in the lab, but the low multitaskers uniformly did better than the frequent multitaskers. Interestingly, those who rarely multitasked thought they did terribly, while the frequent multitaskers thought they performed well!

Despite the evidence, social tolerance for multitasking and its errors is wide. Many states have passed laws outlawing talking on the cell phone while you drive. Not my state, Florida. Only a few Florida counties have restrictions on cell phones and driving. Perhaps we are too entertained and seduced by multitasking to pay attention to its results or to recognize how it can be effectively combated—by doing tasks one by one, in sequence.

The Music of Flow

Mihaly Csikszentmihalyi is not a household name, but the former University of Chicago professor has had more impact than most of us know. The positive psychology movement grew up from Csikszentmihalyi's work especially after it became cross-fertilized with cognitive therapy.

Csikszentmihalyi's understanding of the mind and how we enjoy ourselves is deep and rich. Researching optimal experiences, he eventually came to piece together what is now called flow.

In the deepest flow experiences, people are so engaged in their actions that they forget themselves. Time flies or loses meaning. When you are fully involved in flow, everything you do, every thought and action, follows inexorably and takes up your whole being.

However, many flow activities are much more prosaic. Hopefully (all right, very hopefully) you're having a flow experience reading this book. When people are engaged in reading, they are conversing with the author, checking her examples, creating and internally discussing counterexamples, responding and remembering. Most of us can recall reading a good book and not noticing the time until we were done, feeling exhilaration as the text ended and sadness that there was no more to read. Perhaps the easiest-to-appreciate flow experiences occur when playing games, like tennis. Flow is often involved in playing a sport, dancing, singing, or playing an instrument but can also become part of almost any activity involving work or leisure. Flow adds the possibility of play and creativity to most of the things we do.

Given that flow is so wonderful, it's odd that it can be so readily planned and directed. A few things are needed to set up a flow activity:

1. a goal

2. a challenging set of circumstances

3. skills to meet those circumstances

4. feedback, allowing the skills to be measured and improved

Flow does not involve a constant, nagging look at overall productivity. When you're dancing you may first be concerned about the overall quality of your steps. But as you keep dancing, as you feel the music's rhythms and the actions of your partner, your feet may feel lighter, your steps more accurate and quick, your embrace warmer. When deeply engaged in flow, people often do not notice that time is passing, and they can be shocked at how much time has gone by—or how little. Often their sense of enjoyment becomes most noticeable after the event.

In studies comparing work with leisure activities, Csikszentmihalyi noticed an interesting disconnect. People described frequent and long-lasting flow experiences while they were on their jobs. But once they got home and did leisure activities like watch television, they described themselves as rarely experiencing flow or even real pleasure. Yet many people said that they could not wait to leave work to get home and do the things they "liked," activities like watching TV. Studies done by social psychologist Sonia Lyubomirsky echo these results. Lyubomirsky argues that the greatest factor in personal happiness is not one's relationships but the quality of one's work. Work enjoyment and satisfaction are in turn tied to how much autonomy and control one has on the job.

Csikszentmihalyi, along with many spiritual thinkers, feels that we have considerable control of our consciousness and of how we think, despite the circumstances under which we live. One corollary of this belief is that even boring work can be turned into flow.

Flow on the Job

Over the years I've treated many retired industrial workers. A few of them hated their jobs, but a surprising number found them okay. They enjoyed their coworkers, enjoyed feeling part of a large organization, liked to feel they were in their own way responsible for turning out the machines and products that people enjoyed and needed, whether they were cars or washing machines.

Some would tell me they also had quite a bit of fun on the job. The stories usually had the same theme: these workers were able to turn their tasks into something like a game.

A riveter was required to place the same number of rivets in a circular pattern hundreds of times a day. To make his time flow, he would first rivet counterclockwise. Then he would try riveting clockwise. Often he worked to music playing in his head, riveting to the beat. On other occasions he might try performing his highly repetitive task a bit faster, then take the saved time to more musically perform the next set. Many other clinicians and researchers have seen and written about similar people and situations.

Like many doctors, I hate paperwork. My office manager helps out mightily, but some things I must do. There are days when work feels like a long succession of forms.

I try to leave the most maddening insurance paper shuffles to biological clock dead zones like the midafternoon, when my brain is naturally slower. Though there's no radio in my office, I often work to the music in my head, which also gives me a good idea of my overall mood. If I hear lots of Mozart playing up there, things are going well.

Often I'm reviewing reports I've dictated, which in theory is nothing more than proofreading. However, every medical chart is a potential legal minefield. Things I wrote ten years ago may be resurrected in a divorce case or be used to keep someone from getting disability insurance. Then there are the nagging worries of malpractice—have I mentioned all forty-six side effects this stimulant for narcolepsy might produce, just like the malpractice attorneys told me to?

As I try to look for all these potentially explosive mines, I think about flow. What is my goal? Usually to get through the paperwork as quickly as possible. What's the challenge? To do it skillfully enough that the information I need to impart is communicated without telling so much that it will compromise privacy. What are the skills required? All I know about medical care plus the power to imagine how these records might be used by unknown nefarious persons in the unknowable future. Past depredations by insurance companies and lawyers quickly come to mind.

Suddenly a boring task is not so dull. This man's drug history needs to be mentioned, but not in a way that might hurt his career. Another's medical history is amazingly complicated but there's probably some way I can get it all down in not-too-cryptic medical shorthand that will distill the main facts. Later on, sometimes in unexpected ways, I'll get feedback from other physicians on what I wrote. And since I dictate most chart notes in one take, there are plenty of opportunities with these reviews to improve infelicitous phrases from garbled medspeak into something perhaps more closely resembling English.

When tackling your own boring tasks, keep in mind both the need for skill improvement and the importance of your work. Don't forget that most job tasks are in some way challenging. If you're a customer service representative, you have learned to deal with people who are pleasant, vicious, insane, and everything in between. You have gained major skills, skills you can translate to your home and social life. If your call supervisor demands you cut your call length, you can ask colleagues what phrases they use to get to the nitty-gritty quickest and to solve customers' problems fast. Because it involves goals, challenges, skills, and feedback, flow means you are always learning, potentially improving, modifying what you do with the next go-round. Which is not a bad summation of how the brain works.

Rest as Flow

As you've already guessed, rest is also a set of skills, with goals, challenges, and feedback. You learn and modify these skills as you perform them. Rest techniques provide you many ways to achieve flow.

You have learned many rest skills through reading this book. Consider the first physical rest technique, deep breathing. The more deep breathing you do, the easier it gets. Your breathing becomes deeper and more regular, the sense of rest and relaxation stronger and more encompassing

Techniques of spiritual and mental rest are flow activities that improve with practice. When you start self-hypnosis, it may feel difficult or even weird, but it can quickly become for you the kind of induction that produces relaxed concentration within seconds and can be achieved almost anywhere. Focusing the eye gets easier, quicker, and more interesting as you do it. Contemplating suchness and moving through time and space are more fun when you try them out in different settings and as you seek deeper goals.

All these different rest techniques can become examples of flow for you, and all can improve with time. Your goals may include creating an internal sense of rest and relaxation that simultaneously rests your body and mind. The challenge is to do these rest techniques in many different places, more easily and more quickly, and with greater pleasure. The skill set is made up of the rest techniques themselves while you gain a sense of accomplishment as you learn to rest anywhere and anytime.

As you get good at different rest techniques, you will come to recognize that your mind need almost never feel bored. Prisoners have meditated in cells where many of the good things in life have been forbidden. You can spiritually rest in a gorgeous springtime meadow *and* on a bus ride through a dingy district of a city, noticing the changing forms of nature around you and the shifting flow of life. In the worst times of your workday, you can breathe deeply or engage in big screen technique, which will give you perspective and the chance to plan what needs be done. The goals, challenges, skills, and feed-back needed for flow are present in virtually everything you do and nearly everywhere you look. And all these skills become both easier and more productive if you put them together during the next hours, days, and months.

Sequencing Flow

The scientific literature on dual-task interference explains that tasks that are habitual, well defined, and well practiced need not horribly interfere with each other. Thus we can walk and talk, read and chew gum at the same time. Yet for most definable tasks, doing two things at once will worsen outcomes and decrease enjoyment.

There is one exception to this rule. One task, or stimulus, appears to enhance performance and pleasure. That stimulus is music. Its useful effects are so common they are now being used medically to treat depression and anxiety with circadian precision.

Life is so rhythmic that music is innate to almost all human beings. Making your day fully musical requires you to make it rhyth-mic as well.

One way you provide rhythm in daily life is through sequence. To avoid the pitfalls of multitasking, which include lowered performance and an overpowering sense of stress, you need to intelligently pattern your actions.

If you're looking at your main priority of the day, try to break it down into doable pieces. Science is the art of the solvable. Sequenc-ing is the art of doing many different tasks *one task at a time. This will also help you achieve flow.*

Let's say your main priority of the day is to write a report and ex-plain it to your boss. If you know it will take you at least thirty minutes to write that report, try to find a way to make those thirty minutes uninterrupted. Don't look at your e-mail. If you have your own office, close the door. Turn off your cell phone and hold your landline calls.

As Gloria Mark has shown, interruptions greatly decrease performance. Yet in many jobs, you're going to be interrupted no matter what. So try to break that report you are writing into pieces that are small enough that if you do get interrupted, you will be able to return to your important task without losing too much previous work and too many of your best thoughts.

To break the job down into parts, you will probably need to outline your report first. An outline is nothing more than a sequence of ideas and actions.

As you go through your job, sequence the day. Visualize your workday (rest at work technique 1) at the start of work. Try to recognize the times of day you work most effectively, and schedule your more creative tasks for those times. In general, creative, imaginative jobs are performed best when you are fully alert. Unless you are a lark or owl, these creative times will be in the mid- to late morning or late afternoon into early evening, though there are some who prefer the first hours of the workday.

In many occupations it's uncommon that people can concentrate intensely for more than two hours to three hours at a time, so take that into account. Set up strategic rest periods throughout the day, as you visualized at the start of the day (rest at work technique 1) or in your morning family meeting (rest at home technique 1). Whenever possible, alternate work activities with rest activities, using that simple, basic rhythm to forge ahead to the beat of your own individual biological clock.

If you sequence your day effectively, you will be able to stay on task more readily, and you can multitask as required. To do multitasking well, sequence your tasks. As part of that sequencing, schedule different rest techniques throughout the day to restore you and keep your energy and pleasure from flagging. With time, and as you become more skilled, you will see how each different rest technique can provide its own form and internal rhythm, which can then act as a musical counterpoint to the forms and rhythms of your work activities.

Boredom may be an enemy, but obtaining a relaxed state of concentration, in which you are deeply engaged in what you're doing yet able to withstand interruption, can become a real friend. You want your work to be planned and sequenced, as rhythmic and musical as you can get it. You want your day to flow.

Summary

Multitasking can be a kind of joy ride, but in the end multitasking usually leads to decreased performance and increased frustration. We can't help it—our brains are built to do one thing at a time. However, using the concept of flow, and sequencing our tasks one by one through the day, we can get a lot done and do it enjoyably, with time in between to rest and restore. Flow involves goals, challenges, the skills we have, and the feedback to improve them. Sequencing flow can make even the most boring workday more rhythmic, musical, and effective.

The goal is to do one thing at a time, but in a sequence of activities. With planning and feedback, we should be able to achieve flow more and more every day.

chapter 10

"I HAVE TO DO IT"

Eliminating the Required; Doing the Essential

Do less. Accomplish more. That is what you will be able to do when you know how to really rest. Yet people tell me they can't rest. They have no time—not just time for rest, but for almost anything.

Life should be rhythmic and musical. There should be time for activity and rest, for work and relationships.

But first there must be time. To manage your time, you have to know how and where you spend it. More important, you want to recognize what has the most meaning for you and what gives you true pleasure. To do this you have to sometimes give up doing the required and emphasize doing what is truly essential.

In this chapter you'll learn about some simple but powerful time management techniques and how to apply them. The calculation you will make, however, will be different from that of many books that emphasize time management. Our goal is not to maximize income, though that may happen, but to maximize internal balance and health. You want to increase your effectiveness to do whatever you like to do, creating a clearer focus and rhythm to your life, which then becomes its own reward.

You want to spend your life doing what you care about, and doing it well. Let's take a quick look at the difference between effectiveness and efficiency.

Time Management and the 80–20 Rule

When I was a resident at Bellevue Hospital in New York City I first heard of the 80–20 rule—that 80 percent of the workers did 20 percent of the work. It certainly seemed to fit my work experiences.

In those days, New York City had gone bankrupt. Federal help eventually came to the city and its wholly owned Health and Hospitals Corporation, but for years working at a municipal hospital like Bellevue possessed a Wild West quality. You came to routinely expect the unexpected because quite often you really did not know what would happen next. Someone might come in, take a pistol, and start shooting up the emergency room, or you might find yourself interviewing a patient who had been a resident in an Arizona state hospital for a decade, then put on a bus with a sandwich and a ticket to New York and told to find on arrival a New York City cop who would transport them to Bellevue Hospital.

When the problems in the ER became ridiculous and you really needed help, you soon figured out where to go. Many of the administrators were missing in action, disappearing from their offices into endless meetings or just never showing up. They were especially sparse during night shifts. We soon discovered that one, Mr. Mathura, could be counted on to feel our outrage, roll up his sleeves, and try to get the system working again.

Laboring on the medical wards was not very different. Too many of the nurses and staff were time servers, checking the days to retirement and working hardest at appearing to work. Those who were concerned often ended up doing their own and their colleagues' jobs. That's when I learned about the 80–20 rule, and I watched it in action day by day.

In later years I learned that the 80–20 rule possessed a very long history but stemmed in part from work done in the early twentieth century by the social philosopher and economist Vilfredo Pareto. Pareto pointed out that for economic results, frequently 20 percent of the investment creates 80 percent of the profit. The rate certainly varies, following a simple mathematical formula, sometimes hitting

90–10, other times 75–25, but the point is clear—a minority of your effort and time produces the majority of desired results.

Timothy Ferris in his fascinating *Four-Hour Work Week* has popularized Pareto's formulation, known as the Pareto Optimum. It's part of his program to let people outsource many of their tasks in order to spend fewer and fewer hours on whatever job they work. The goal is to join the ranks of the "New Rich," self-motivated people who define their lives through active forms of leisure and work that they personally control.

Ferris's book provides lots of useful techniques, but he has done another service by adding Pareto's insight to the too-often-neglected rule named Parkinson's Law. In the 1950s C. Northcote Parkinson made this astute observation: activity, particularly work, will fill up the time allocated to it. If you have four days to write a report, don't be surprised if it takes you the whole four days.

The Pareto-Parkinson Optimum combines two powerful facts— that short periods of directed effort usually get most of the work done, and that given the chance, people will fill up all the time they have with whatever task is at hand. Not surprisingly, people often discover themselves watching television show after television show when the workday is over unless they have clear goals they want to accomplish and plans to accomplish them.

However, the Pareto-Parkinson Optimum is usually applied to business outcomes that can be measured by clear fiscal metrics. For those outcomes it is easy to see the result: either you make money or you don't; you close the sale inside three hours or you fail. A more interesting use of the Pareto-Parkinson Optimum is to apply it in areas not so easy to quantify—love and work.

Love and Work

When Freud was asked what the two most important things in life were, he gave a succinct answer: love and work. I think he was right.

Love is of course defined far more broadly than romantic love. Love involves all the relationships in your life. Generally the most important relationships are with other people: parents, siblings, spouses, children, friends, and acquaintances. However, there are many other types of relationships. Some people feel closer to their pets than to many of their relatives. Others feel that their relationship to nature, to God, to the spiritual, is as important to them as their social relationships.

Similarly, work is something much larger than what provides you a paycheck. Work also involves relationships—with individuals as well as ideas. A mother or father who homeschools a child is certainly working, as is a Little League coach training her team or an executive mentoring a young entrepreneur.

When you consider that love and work both involve many different kinds of social relationships, the distinctions between love and work suddenly appear thinner than we usually think. What truly places them together, and is worthy of its own Pareto-Parkinson calculation, is the issue of personal meaning. What really counts in your life?

Consider a relative of mine, a likable man who has practiced medicine for thirty-five years despite many health reversals. While I am writing this chapter, he is covering the patients of two of his colleagues. One colleague is sixty-five years old and suffering from a tumor metastatic to his brain; the other is seventy and living with lymphoma, trying to practice in between bouts of chemotherapy.

Both could quit working. For them money is not the defining issue. They like their work. They want to practice medicine—until they drop. Otherwise, they claim, they would lose much of their sense of individual purpose and sorely miss the relationships they have built up with their patients. In their minds, love and work meld.

In this way, these men are fortunate. Some people regard work as a required form of financial slavery, something they would dispense with in a heartbeat if they won the lottery or an old uncle provided them a sizable bequest. Others prize their leisure time above all, finding watching TV or electronic entertainments the desired peak experiences of their daily lives.

As we noted in chapter 9, when people rate their activities at the times they are actually doing them, a different picture emerges. Csikszentmihalyi found that people were often disinterested and unattached while engaged in the leisure activities they could not wait to leave work for, yet they were far more prone to describe themselves as experiencing flow while at work. Many denigrated work compared to leisure pursuits though they actually enjoyed work far more while they were doing it.

Social expectations change what we think, just as they influence what we do. For many, the successful individuals are the ones with wads of money who can pursue any leisure activity they want anytime they wish. By contrast, I've learned from treating wealthy businessmen who retired early to live in sunny Florida that nearly unlimited time for leisure pursuits does not automatically confer either happiness or meaning. Many find that meaning, purpose, and the strength of their relationships are more useful to them than their money. To test if that is true, let's take a brief quiz to find out what Pareto and Parkinson might have to say about your life.

Listed below are the twenty-four hours of a typical workday. For most of us, that day will take place during the workweek rather than on the weekend.

Please note your activities, hour by hour, during that typical day. Then rate each hour on a scale of 1 to 10, looking at how meaningful you found the activity and how enjoyable it was.

For meaning, rank as a 10 the moment in your life when you felt most proud of yourself and your life's purpose. For enjoyment, rank as a 10 the peak experiences of your life, when you felt better than you ever had before. For meaning, rank as low (1–2) periods when you felt truly ashamed, and for low levels of enjoyment, times when you experienced uncontrollable distress.

WORKDAY

	Activities	Meaning (1–10)	Enjoyment (1–10)
Midnight–1:00 a.m.			
1:00–2:00 a.m.			
2:00–3:00 a.m.			
3:00–4:00 a.m.			
4:00–5:00 a.m.			
5:00–6:00 a.m.			
6:00–7:00 a.m.			
7:00–8:00 a.m.			
8:00–9:00 a.m.			
9:00–10:00 a.m.			
10:00–11:00 a.m.			
11:00 a.m.–Noon			
Noon–1:00 p.m.			
1:00–2:00 p.m.			
2:00–3:00 p.m.			
3:00–4:00 p.m.			
4:00–5:00 p.m.			
5:00–6:00 p.m.			
6:00–7:00 p.m.			
7:00–8:00 p.m.			
8:00–9:00 p.m.			
9:00–10:00 p.m.			
10:00–11:00 p.m.			
11:00 p.m.–Midnight			

As they do this quiz, people often point out to me that they were asleep for seven to nine hours every day. How can this rank as "activity"?

Simple. As you learned throughout this book, rest restores you physiologically, mentally, and psychologically. I know many people who regard sleep as one of the most enjoyable things they do.

Sleep is also meaningful or, rather, full of meaning. Proper sleep is critical to laying down your memories, perhaps the most precious commodity you own as you move through life. Sleep is necessary to learn well. Sleep is required to control your weight and improve your mood. Sleep makes people feel rested, comfortable, and whole. All these things that sleep does are meaningful

Other people find the hour slots of this quiz too much of a straitjacket. They describe their work activities as so varied they can't put them all down inside a single hour. If that's the case with you, break up your work hours in half-hour segments and rate each of them.

Now perform the same quiz, hour by hour, for a typical day that is not a workday. Generally this day will take place during the weekend.

NONWORKDAY

	Activities	Meaning (1–10)	Enjoyment (1–10)
Midnight–1:00 a.m.			
1:00–2:00 a.m.			
2:00–3:00 a.m.			
3:00–4:00 a.m.			
4:00–5:00 a.m.			
5:00–6:00 a.m.			
6:00–7:00 a.m.			
7:00–8:00 a.m.			
8:00–9:00 a.m.			

	Activities	Meaning (1–10)	Enjoyment (1–10)
9:00–10:00 a.m.			
10:00–11:00 a.m.			
11:00 a.m.–Noon			
Noon–1:00 p.m.			
1:00–2:00 p.m.			
2:00–3:00 p.m.			
3:00–4:00 p.m.			
4:00–5:00 p.m.			
5:00–6:00 p.m.			
6:00–7:00 p.m.			
7:00–8:00 p.m.			
8:00–9:00 p.m.			
9:00–10:00 p.m.			
10:00–11:00 p.m.			
11:00 p.m.–Midnight			

Now take stock.

What did you rank as the two most meaningful hours of the workday?

The two most meaningful hours of the nonworkday?

What were your activities during the two most meaningless waking hours of the workday?

What were you doing during the two most meaningless waking hours of the nonworkday?

What were you doing the two most enjoyable hours of the workday?

What were you doing the two least enjoyable hours of the nonworkday?

Check your answers, and think about them for a few minutes. Write down the activities you found the most meaningful here:

_____

Put the most enjoyable activities here: _____

_____

Looking at your workday, what was the single most meaningful hour? Now look at your most enjoyable hour.

If they are the same hour, you're fortunate. If they're not, ask yourself why. Are meaning and enjoyment quite distinct for you?

Now, look at how much of your time was actually spent in the most meaningful and enjoyable activities. Was it 20 percent of the day you worked, or was it less? If the figure is well above 20 percent, count yourself fortunate again.

Using Pareto-Parkinson Optimums for Workdays

Now you will try to appropriate Pareto and Parkinson to your own cause. Look at those activities you find meaningful and enjoyable. Now see what you can do to increase the hours you spend on them.

For the waking activities you dislike (just the waking activities; people are often too prejudiced against sleep to give it its proper requirement), apply Parkinson's Law. Try to give those activities the *least* amount of your time.

Hate writing up reports? Doing paperwork? Then cut the time you spend on disliked activities by thirty minutes each week. Keep cutting for a month, and see if you can properly finish them all anyway. Cut the time until you can't cut it anymore. Don't be surprised if you find several hours of your week are now free. Allocate a decent part of the time saved to active rest techniques.

Unfortunately, you may be in a job where your autonomy is limited. Your boss demands you spend time on jobs you find meaningless or silly, and that's that.

In that case, immediately apply principles of flow to the work you dislike. One of my patients is a postal employee. Due to personality difficulties with her supervisors, she often finds herself stuck in customer service roles when she'd much rather be out in her truck delivering mail.

She likes delivering the mail. She recognizes its usefulness to society, and she finds that part of her life meaningful. However, dealing with the Great American Public is not her cup of tea.

Still, she tries to make her hours at the front desk as interesting, pleasant, and useful as she can. As she fills out forms and weighs packages, she talks with the postal clients, trying to find out what they really need. Then she tries to meet those needs. Gregarious by nature, she jokes as she works. Her good humor appears to drive her supervisors nuts, but her attitude allows her to create brief social connections that make the job feel more meaningful—and make it more enjoyable.

Pareto-Parkinson Optimums for the Nonworkday

Now study your quiz for the nonworkday. Look at the most and the least meaningful hours.

Did you spend the majority of your time on activities you found meaningful? Or did your nonworkday also fall into a standard Pareto distribution, where the large majority of your time was spent on things you don't find important to you?

Do the same calculations with your most enjoyable active hour. Is this activity something you get to do a lot? Or is it the small sliver or reward you allocate yourself when you think you've got enough time to do it?

Thinking of Pareto and Parkinson, what can you do to increase the amount of time spent on things meaningful and enjoyable? What can you do to decrease the time spent on things you dislike?

People who have done this exercise are frequently surprised by the results. They discover they are spending weekends ferrying their children from play date to play date, sports practice to sports practice, or class to class. They really dislike all the time they're spending in a car and the fact that they end up seeing their children far less, and with less intensity, than they like.

I tell them to spend more of the time with their kids directly and conjointly. Rather than cleaning a child's room alone, do it with them. Rather than clearing the plates after a meal, have everyone in the family help out.

If your kid is simultaneously engaged in three sports, see how much they love them before spending that much of your "free time" in a car. Few children will grow up to become successful professional athletes. They might have more fun riding a bike alongside you.

Other people are surprised to find how much time they spend in front of the TV. They say they enjoy watching TV, but rarely do they consider it meaningful. I tell them to try different activities, like calling up a friend or reading a book on a subject they wish to learn. I also tell them to always apply the techniques of active rest, especially social connections, to make their time more enjoyable, meaningful, and pleasurable.

Survival, Meaning, and Enjoyment

There are times when getting by can be really rough. Paying the bills and keeping the household going can take most of your time and most or all of your effort.

It's when you feel you've hit the wall that issues of meaning and enjoyment really come to the fore. If you're working seventy hours a week to keep a home you can't afford, then, if the market allows, consider a different home. Studies of McMansions find that people occupying 10,000 square feet of space are often spending virtually all of their time within 1,100 square feet. It's another example of Pareto's Optimum, except in this example applied to space, not time. For lots of things E. F. Schumacher was right: small, or at least smaller, can be beautiful.

Others have reasons to work endless hours. I often hear immigrant parents say that they want their children to have more opportunities than they had. Even if they don't enjoy their work much, they find the results of their jobs quite meaningful.

Such decisions are very personal. Making the right calculation is something you'll have to do for yourself. Fortunately, there are useful ways to measure that balance.

Flow and Complexity

Look at the periods you spent doing things you felt enjoyable and meaningful. Chances are that most of those times have a strong sense of flow, of moving forward and challenging yourself.

Flow activities do more than give you a sense of engagement and excitement. When you have set goals, work to those challenges, use your skills, and watch them improve, you change as an individual.

As you do more and more flow activities, you as a person will become more interesting and your life will become richer. As you understand what you care about and do more of it, balancing activity and rest, experiencing more flow in your life, your allotment of both meaning and enjoyment should grow.

As you become more interesting to yourself and others, you will recognize that you know more and have learned more. Not only are you more skilled, but by learning and doing you have become more helpful to other people.

You may also become more understanding. As you improve at different activities you understand more of what you like and what you don't like. You see what you're good at *and* what you care about. As you understand yourself, it becomes easier to understand others and their motivations and concerns.

The more you engage in flow activities, the more you grow. You find that you can be creative in areas that you might previously have thought trivial, finding new ways to perform the simplest tasks. As you do more, you become more productive.

Though many people think they prize their simple passive activities, such activities are rarely very meaningful or pleasurable in the long run. As people learn more, they see more, understand more, and create more themselves. They develop new capacities and more confidence, which allows them to appreciate the world more fully. They also learn to appreciate rest.

Rest and the Essentials

Supporting and increasing your ability to love and work can make for a better life. You also attain a better sense of balance. Rest is as critical to life as activity. Rest aids your survival, your pleasure, and your ability to find meaning in the world.

Like other flow activities, rest skills improve as you do them. As you actively rest, performing the different techniques of physical, mental, social, and spiritual rest, you may see how interesting the ordinary world can be.

Spiritual rest can connect you with a sense of meaning quickly and powerfully. Social rest can make even seemingly pointless encounters enjoyable and enriching. Mental rest can take you out of your individual space and show you the way to find peace and meaning in the small details of life.

Summary

Too much of life can be spent doing things you regard as required but not desired. In this chapter you've learned to begin to look at what you consider meaningful and enjoyable in your daily life, and how to increase both. You've looked at how to rebalance love and work, and how to use active rest techniques to obtain that balance.

Meaning is essential in life. Rest is also essential. Together they create a life that is powerfully balanced.

chapter 11

TUNING YOUR LIFE

Setting a Rhythm for Your Day

I tell people they can make their lives become musical. They say I must be kidding. "How can I keep my job, take care of the kids, pay attention to my spouse so he won't leave me, pay the mortgage and taxes, visit my mother every Sunday and have my life flow like music?"

I hear such comments often. I also hear people say they can't rest, they can't do these quick rest techniques, there's just no time for any of that stuff. And they feel there is certainly no chance that they can make their lives rhythmic, let alone musical.

Let me prove to you that you can make life musical. All you need to do is walk.

Music and Rhythm

There's much more to rhythm than just rhythm. Cognitive neuroscientist Daniel J. Levitin began his career as a record producer. Then he wanted to find out what music physiologically did to people. In his excellent *This Is Your Brain on Music* Levitin explains the major elements of music, one of which is rhythm. Yet rhythm is actually three things—rhythm, tempo, and meter. What exactly are those three parts?

Rhythm is the time pattern of notes—long and short.

Tempo is pace—fast or slow.

Meter is beat—what beat is emphasized and which beats are not; downbeats and backbeats.

Now it's time for you to personally discover these three parts of rhythm using your own two feet. Find a nice, comfortable place to move, perhaps the carpet in front of the seat where you're reading this, and take a short, normally paced walk. Notice how long or short your paces are.

That's your walking rhythm. If you're like me, right-handed and right-legged, you'll see there's a small difference in how long it takes to march with your right foot as compared to your left.

I push off more with my right leg and generally have a very slightly longer stride with my right leg than my left. Perhaps your motion is perfectly symmetric and takes the same time for each side.

Now take a long stride. Try it again. Follow it with a short stride. Do long followed by short strides a couple more times.

You now know what rhythm is—the pattern of long and short strides. But rhythm also involves much more than how we move.

The next part of rhythm is tempo. Again, walk a short while at your normal pace. As we have discussed several times in this book, your normal walking pace will probably be different, often just subtly different, from that of almost anyone else—your brothers and sisters, your parents, your boyfriend or girlfriend. When we go out and walk with others we generally adjust to the slower pace around us. For example, if you walk with an older parent, you will probably shift your tempo down. That tempo is the rate at which you stride.

Tempo has a lot to do with how we move and emotionally feel. Generally, if we're not fleeing a mugger or a long-lost, blissfully forgotten junior high school classmate, moving fast makes us feel a little happier. If we move more slowly, much slower than our natural walking pace, we might begin to feel tired and a bit blue. Tempo changes the feeling of life. Virtually every activity we engage in has its own specific tempo—the rate and speed with which we act.

Meter is really a matter of emphasis: which beat is stronger, which weaker. Whenever you walk you have a natural downbeat, rolling from your heel and pushing the front of your foot onto the ground, followed by a physically literal upbeat, in which you lift your foot from the earth. If you follow your steps closely, you may find that you probably have a stronger downbeat if you are moving down on your dominant foot (90 percent of us are right-handed though not necessarily right-legged) as compared to your nondominant foot.

Try another simple, brief walk. Feel the downbeat? Still can't? Then return to walking to music (mental rest technique 3) Many of the songs or melodies you listen to will have a strong four-beat meter, with the downbeat on the first beat and the other three beats getting less emphasis.

It's just like the title character of Molière's great play *Le bourgeois gentilhomme*. The bourgeois gentleman is overwhelmed to learn that he is speaking prose. In a similar way, music is a part of our everyday actions that we too often overlook. If you can't hear the rhythm in your own footsteps, listen to the footsteps of someone you know—or even someone you don't.

Remember those 1930s movies where the hero or heroine identifies the perpetrator by the way he or she walks? Take a few moments and listen to the walking of someone you know well, like a parent or a child. Chances are you can identify the person by the rhythm, tempo, and meter of their walk, even with your eyes closed. Chances are also pretty good that you can also identify their rhythm, tempo, and meter when they start to run.

Now, if you are taking a pleasant walk amid greenery, as in walking in a park with a friend (social rest technique 5), take a moment to listen to the movements of the people approaching and leaving you. There is the shuffle of the old; the skipping of children; the slow, sweet gracefulness of a young woman who has just seen her lover; and the brisk, measured tap of the businessman who knows he is late for a meeting. If you listen more closely, you'll probably find that most people have their own characteristic movements and sound as they walk, as individual to them as their voice and their smile. There's much music and rhythm throughout human life, for the simple reason that we're built that way.

Music, Rhythm, and the Brain

The old Latin saying is true—time rules life. In our case, time is expressed through rhythms that mirror the natural environment in which we live. Our inner biological clocks allow us to accomplish all our physiological goals from survival to procreation in full, musical fashion.

In this book you've heard a lot about twenty-four-hour rhythms, rhythms that fit the patterns of day and night, sun, moon, and stars. The twenty-four-hour day is a basic design element of terrestrial life.

But there are many more rhythms running through our lives and bodies. Some we know well—the rhythms of the year, annual events like the weekend of July 4 and paying taxes on April 15. There are the rhythms of the seasons, which is why northeasterners tend to get depressed in the winter months, with their lack of light, and start imagining moving to California or Florida. Sometimes they take their winter vacation in the southlands and find the climate and lifestyle make them feel so much better that they *do* move. (Unknown to many, it's the light, not the warmth, that usually improves their winter moods so much.) The monthly rhythms of the moon are enshrined in many basic physiological paths, including our critical procreation cycle, without which our species would not survive. Ovulating at times that fit when the moon is waxing or waning does not make women lunatics. Still, we have named the subsequent basic life process in English with a sadly unmusical term—menstruation.

Yet there are so many more inner rhythms that time our lives. Physiologists have discovered sixty- and ninety-minute rhythms. In people, these hourly and ninety-minute rhythms are so easy to make routine that they eventually enhance insomnia. I wish I had a nickel (with inflation, maybe a quarter) for every patient who told me they wake up at a set hour and then every hour thereafter. They have habituated, "entrained," their hourly internal rhythms to wake them throughout what should be a sleep-filled night. They wake first, then again, hour on the hour, because they have set their insomniac waking pattern by glancing at the clock when they open their eyes (please, try not to look at the clock during sleep hours until you are sure you wish to remain awake thereafter). There are also the shorter rhythms of regular cell interactions, including the minutes to hours of effects produced by different hormones and the effects of nerve cell firings, lasting milliseconds to seconds. Physiologist Rodolfo Llinas of New York University believes the true "primal" rhythm of the body is an internal cellular clock that chimes out at the rate of 1,200 times per second. These many rhythms of the body not only create the clocks that time everything that we do, they are also probably ways our cells communicate. Much of that communication takes place in the brain.

Living Between Energy and Information

If you are a chemist trying to reproduce life in a test tube, your first requirements are deceptively simple. There are two—energy and information. The energy can come from the electron interactions of sulfur-containing molecules, as do the bacteria recently discovered enjoying an icy existence in the airless darkness three hundred and fifty meters below the Antarctic ice shelf, or it can come from sunlight, which in plants activates chlorophyll, a molecule remarkably similar to hemoglobin, which pushes oxygen into our cells so that we may live.

But to use energy to obtain life, you first need information. We store most of it in different kinds of smart molecules, like proteins and glycoproteins, or in the master memory molecules that contain our genetic information, RNA and DNA. It may turn out that DNA alone is not all that is required, in the style of *Jurassic Park*, to reproduce the species of the past or, in sci-fi fashion, to create the new and potentially frightening life forms of the future.

For life to exist, information has to be quickly and efficiently passed from one entity to another. You need a system, separate programs, and one or more languages. The genetic code of DNA is one such program, the neural connections of the brain another. Interacting and intelligible communication is necessary to make life. So how do you pass information? On this planet a fundamental element of biological communication is rhythm. Rhythm is a big piece of how our cells communicate and interact.

Francis Crick was the codiscoverer of DNA's structure with James Watson. Once they figured out DNA's structure, they immediately understood how its paired molecules created and reproduced information. Crick was a physicist working in biology, a fish out of water who loved living in pure air and who realized how critical information was to all the life sciences. One of his last major hypotheses was that consciousness begins when nerve cells start firing together at the rate of 40 hertz—40 cycles per second.

If you ever watch the firing of nerve cells, you will realize how profoundly rhythmic it is. Nerve cells communicate through passing—or not passing—information. What matters is firing or not firing, the same message or no-message pattern we see in computer bytes or the dots and dashes of Morse code. From such rhythms come the physiology of vertebrates, the active patterns of speech, the budding science of computers, and what we like to call human civilization. These rhythmic patterns of nerve cells create this sentence and simultaneously produce your ability to read it.

There are perhaps a hundred billion nerve cells in the brain. Each may have ten thousand connections with other nerve cells. We can think of each of those connections as producing a letter in a word. Our English language operates with a mere twenty-six letters. The number of potential interactions of nerve cells approaches infinity. If they operated according to standard biophysics, they might produce cacophony. But they don't. They produce thought and speech, prayer and love. When you listen to the firing of nerve cells in a laboratory, you do not hear pure noise. What you are listening to is new information. Right now tens of thousands of nerve cells will fire in unison as your eyes pass over this page, and probably even more when they focus over this small dot.

Firing or not firing, firing in unison or alone, firing in patterns that combine and link with other patterns—the electrical activity of brain cells has its own crazy music. When you study deep sleep in animals and witness their nerve-firing patterns, you often see the same patterns of activation and inactivation that occurred during the day. It's like watching a TV program rerun—there they are, the same daytime firings and fusions of pattern that were seen when the rat learned its way through a maze or the mouse inched its way to a piece of food. In sleep, these patterns of nerve activation and connection are alternately weakened or strengthened. We call the result memory. It is through the rhythmic firings of our nerve cells and the biochemical changes they create that we learn and we remember. Not incidentally, it is also how we experience joy.

Flow and Music

For many people, flow experiences are also peak experiences. When people are deeply in flow, they feel so engaged they don't notice anything else. Generally they do not notice or care about the passing of time. All their attention moves to what they are doing—the challenge, the use of their skills, the sense of immediate engagement tempered by discovering whether what they are doing is effective or not.

Flow activities are generally rhythmic. The classic flow experience for many of us is playing a game like tennis or soccer. In tennis there is the back-and-forth thwack of the ball, the intense speed of the first serve and the slow arc of the lob. In soccer the flow of the game can literally be marked across the field, as the players follow the ball or position themselves to be where they hope it will arrive, stopping and starting, running and skidding, racing to kick or suddenly stalling to put off the rhythm of another player. Whatever game people play, they can quickly tell you whether they are in "the groove," playing with the intensity, smoothness, and musical rhythm that mark our best efforts.

This same term, *in the groove,* is used by engrossed musicians, neurosurgeons in the middle of successful operations, and actors happy with their performance on the stage. We know when we are really musical and when we are not. Thankfully, we can use the concept of flow and rest techniques we've detailed in this book to make most of our days more musical.

Some flow activities, like reading a good book, even this book, may not appear to be immediately musical. But if you look at what reading involves, you may notice that language itself is patterned in very specific, highly rhythmic ways. The greatest writers, such as Shakespeare or Dante, write in rhythms so obvious and idiosyncratic we can often identify the author by the metrical pattern of the words alone. The rhythm of Dickens is completely different from the rhythm of Hemingway, and we enjoy these musical aspects of books much as we enjoy the rhythm of our favorite songs.

Throughout this book you have already learned, perhaps without fully recognizing it, many different musical ways of resting. Performing deep breathing (physical rest technique 1), whether it's to the count of four while breathing in and eight while breathing out, is a rhythmic experience. Add to it fully sensing the feel and flow of air through your lips and mouth, down your pharynx, into your trachea and the gorgeous spheres of the alveoli in your lungs, and it may soon become a restful, musical, flow experience.

You can then extend that musicality by deep breathing while doing mountain pose (physical rest technique 2). Feel the rhythm— breathing in to the count of four, the slow breaths out to the count of eight. Feel the tempo, the rate, of each breath. You have many choices. You can count quickly, one-two-three-four, or slowly, one one thousand, two one thousand, three one thousand, four one thousand. The slower rate may provide you a very different, perhaps more restful, experience. When in mountain pose you can feel the internal meter, the sharp downbeat of breathing in, the more measured, gradual backbeat of exhalation. You quickly come to see that you are using the rhythms and forms of music in the simplest act, that of breathing.

Once you've breathed musically, sensing how your breathing rhythm feels throughout your body, you can then see how musical rhythms may be applied to most of the rest techniques in this book. When you walk in a garden with a friend, you can walk to music, with both of you moving to the same tune. When you do self-hypnosis, you can speed up or slow down your thoughts to the rhythm of your breaths, fitting those thoughts to the inherent musical rhythms in your body. When you perform the techniques of spiritual rest, you can also do them rhythmically. For example, when moving through time and space (spiritual rest techniques 2 and 3), you can visualize the patterns running at the pace you decide—quick, superfast, or slow, working to any tempo you like. You can use these varied rest techniques together or apart, fitting them together in ways that will over time feel more musical to you.

Or you can use music directly.

Using Music to Make Life Musical

The development of the transistor changed human life. You could put people into space and keep them there, shrink computers from factory size down to nearly invisible silicon chips, turn phones into powerful mobile computers. You could also take your favorite music with you everywhere, from your bedroom to the garage, schoolyard, park, or mall.

Transistor radios seem quaint these days. The Walkman demonstrated through its name that music and mobility could become a unity, and the iPod and its variants allow you to take a giant musical library with you onto a sailboat becalmed in the Pacific or on a trek through the supreme heat of the North African desert. Music, all music, can now go anywhere—even to places you can't go.

As the brain operates musically, it's no surprise that people like adding music to almost any activity. In the subway or a transoceanic plane, many people do not want to give up listening to their music for a second, and certainly not enough to pay full attention to their fellow passengers.

Please use music to set a musical pattern to your activities or as something to which you will listen intensely. Please don't try to do both. Someone like me, trained in childhood to listen to classical music, finds polyphony so engaging it's hard to do anything but listen to the music. That's a problem when you're driving or when you hear distracting background music while talking to a colleague. Intensely listening to music can be thrilling, exciting, and powerfully pleasurable, but it can also create the conundrums of multitasking and can prevent you from performing necessary or vital tasks. I've watched many pedestrians, cocooned inside their ear pieces, walk across the street oblivious to cars speeding in their direction. I've watched people like that get hit and dragged across the road. You don't forget it.

Instrumental and vocal music are great things in our lives, but try to use them wisely. Some writers type most happily to Bach while others find themselves hopelessly stopped in their tracks by the first three or four notes of a familiar melody. Each of us uses music differently, often for different tasks at different hours of the day. In many cases it's fun to use music to set the pace for your tasks, but please focus exclusively on music only when it is safe to do so. For example, in walking to music (mental rest technique 3), the goal is walking, moving rhythmically, enjoying the feel of your body as you move, concentrating on the dancing feel of your walk as well as safely reaching your ultimate destination. Music can provide wonderful rhythms for many of our actions, but you also need to remain conscious of what you are trying to get done.

Full, attentive consciousness is important if you want to fully achieve active rest. The active rest techniques you've learned are goal directed, and among those goals is feeling more rested, more alert, more alive, and more centered. Music has great power to improve our attention and increase our ability to focus and be aware because it is so close to how our brains actually work.

So use music often, in short spurts and long, inside and outdoors, all the days you wish. The music can come from an iPod or a cell phone, a TV set or a computer, or from your own head. It can also come from the rhythms of your daily life.

Tuning Your Life by Going FAR—Using Food, Activity, and Rest

Music can be a little complicated to analyze. There is rhythm, tempo, and meter. There is pitch (rising and falling sounds), timbre (the quality of sound; the difference between a note played on a trumpet or an oboe), melody (specific pitches, organized rising and falling patterns that fit what the brain enjoys), key (simultaneous pitches that fit culturally conditioned brain expectations), harmony (keys and melodies fitted together), and many other yet more complex patterns. Fortunately, music is so much an innate part of us that we can get fairly far in making our lives musical just by using rhythm alone.

A basic rhythm of life is food, activity, and rest. If you watch animals in the wild go through their daily lives, most will first eat, then move, and then rest. There are, of course, exceptions. Some big cats like lions or tigers that need not fear predators (that is, except for us) sometimes operate to the rhythm of eat, rest, and move. The same is true of domestic cats and dogs.

If you watch human children, however, the human pattern of food, activity, and rest is pretty obvious. Kids eat and play, *then* they rest. They also move rhythmically—and as a group.

Play a song, and kids will begin to move. Not only will they move, often to their own dance steps, but they will move together and to the same rhythm. This combination of music and motion is enshrined in many human languages. The word *rhythm* is attached to original meanings of measure, movement, and flowing water. As Oliver Sacks points out, though in English we separate the words *song* and *dance*, in many different tongues one word is used to signify both.

We should not be surprised. Listening in padded seats inside a concert hall to professionally trained musicians is an anomaly of modern life. In most nonindustrial societies, almost everybody can sing and dance, and most are expected to do so, brilliantly or not. Despite our work stations and desks, televisions and monitors, cars and trains, our bodies are not built to sit still. They're built to move, and to move musically.

Going FAR—setting life to the rhythm of food, activity, and rest— has many advantages. In the rest of this chapter we outline a few of them.

1. *FAR is easy to conceptualize.* All you need do is remember three letters and what they stand for: food, activity, and rest.

2. *Going FAR can help you control your weight.* Most of us have been taught since childhood to sit and slowly digest our meals. It's good advice if you have a Depression-era population that can't get enough calories. It's particularly good advice in places where people are starving. But it's bad advice for a population that produces nearly 4,000 calories of food a day and needs only 2,000 to 2,300 calories to survive and be healthy.

Our food, especially the standard American diet of starch-, fat-, and sugar-laden fast and processed foods, is digested very quickly. However, we possess a thirty-foot gut for a reason. Most of the food we evolved with was fibrous and hard to digest. Dogs have a six-foot gut because dogs are carnivores. Feeding dogs meat won't cause their arteries to fill up with plaque. Give us fatty meats to eat throughout childhood, and we'll develop atherosclerosis faster than you can say "middle age."

Yet there are advantages to slowly digesting food. Fibrous foods, the sort we evolved with over the last couple of million years, take a lot of work to digest. The sugars they possess are pried apart and absorbed rather slowly. As a result, glucose levels don't rise rapidly.

Because glucose levels do not rise quickly, you don't get huge insulin peaks. Because there are less intense insulin peaks, you don't store so much fat in your abdomen (the excess sugar has to go somewhere, you know). Fat in your abdomen, especially fat stored around your internal (visceral) organs, is now recognized as a major hormonal gland, producing all kinds of potentially bad stuff. More hormones will probably be identified as originating in visceral abdominal fat. The end result of many of these rapidly secreting abdominal fat cells is more plaque plaguing your arteries.

Fortunately, this adverse cycle can be short-circuited by doing something really simple—moving after you eat. When people move, the gut more or less shuts down. The blood flow gets switched from the splanchnic circulation, which courses through your guts and abdominal organs like the liver, to your outside muscles. The blood supply shifts to the muscles that allow you to move.

So when you're moving after the meal, you won't digest food quickly or perhaps will temporarily stop digesting altogether. You won't get those big glucose peaks. You won't overtax your insulin-producing pancreatic islet cells, a process that leads eventually to diabetes and real trouble. You won't get huge insulin peaks followed by hypoglycemic episodes that make you even more hungry—one of the reasons McDonald's likes you to finish your meal with a dessert. Without those huge insulin peaks, you probably also won't get such big fatty deposits building up in your abdomen.

In other words, moving after meals can help most of us lose weight. In my experience, if I can get people to stroll or walk after each meal, they generally lose around fifteen pounds in two to eight months. They also get smaller waists. That is itself a major health issue, as waist size is a better predictor of survival than your weight. Moving in sunlight, which activates high-energy-using brown fat in animals, might also help you slim your waistline.

3. *Going FAR helps prevent gastroesophageal reflux.* Gastroesophageal reflux disease (GERD) affects hundreds of millions of people. Stomach acid moves up through the lower esophageal sphincter and

attacks the acid-sensitive cells of the esophageal lining. Over time, it changes the types of cells lining the gut. With enough time, some of those cells turn cancerous. Esophageal cancer does not have great survival rates.

Going FAR curbs the process. It's simple physics—a matter of gravity. If you stand up after a meal, gravity helps pull food down into the stomach and gut and makes it harder for acid to travel upward.

Before you get your gastroenterologist to fetch her endoscopes and perform the bowel run of the stars, looking for erosions and tumors throughout your gut, stand up and walk after you eat. It should prevent up to 50 percent of GERD besides helping you control your weight.

4. *Going FAR can improve your mood.* Sunlight resets biological clocks. It changes the nature of immunity, provoking production of natural killer cells and setting up many of us to more effectively fight infections. It creates vitamin D, which is at the moment becoming heavily promoted as a cancer preventer.

Sunlight also makes you happy.

I'm one of the 25 to 50 percent of North Americans who find the winter months in the northern part of the country chilly in terms of both temperature and emotions. When I first moved to the Sunbelt, to do a medical internship at the University of California–San Diego, my near complete lack of free time did not hide from me the fact that my mood was much better in the winter months on those rare days I got outside and saw sunlight—or just stayed close to windows. Teaching at the University of Texas–Houston reinforced the message that light changes one's mood. Suddenly winter was a great time of year. I've lived most of my life in the Sunbelt ever since.

When you move after meals, even if it is during a Wisconsin winter, you may improve your mood. In overcast Basel, Switzerland, seasonal depression has been successfully treated by getting people to walk outside under heavily overcast skies. Even cloudy days can improve your mood if you are out there in the light.

Physical activity itself also improves mood. For people who are mildly to moderately depressed, physical activity, particularly walking, helps improve their mood as much as antidepressant drugs do. Some people can get over even severe depression through high levels of aerobic physical activity.

5. *Going FAR improves your health.* Moving after meals is exercise. Anything that engages your major muscle groups is exercise.

Don't believe it? In the European Breast Cancer Trials, one great way to decrease breast cancer risk was to do housework. The more housework done, the lower the risk of breast cancer.

Even your attitude about exercise makes a difference. One entertaining Harvard study from 2007 looked at female maids in Boston. All of them did the same work, in much the same way. They were split into two groups. One group was told that what they did every day was just fine, please keep it up. The other group was told that their work was exercise and by itself fulfilled the surgeon general's guidelines for the amounts of daily exercise needed to achieve better health.

What happened? Members of the group told they were performing exercise while they worked dropped their weight, many of them quite a few pounds. Not so with the control group. The group told they were exercising also notably decreased their cholesterol levels, which did not happen to the control group. Just letting people know that moving is a form of exercise can itself improve their biological health markers.

How much exercise is optimal? The Canadian government suggests moving your body ninety minutes a day. Many researchers think optimum benefit comes with around sixty minutes a day.

So take a moment and imagine what you can accomplish by walking ten to fifteen minutes after each meal. Thinner waistlines, smaller weight, better lipid levels. Besides reducing your risk of cancer, you'll probably get better-looking skin. All from adding a little rhythm to your daily life.

6. *Going FAR can improve your social support.* There's no need to eat alone. We are very much social creatures.

You can go FAR to create an armature of food, activity, and rest throughout the day. It's generally much more fun to eat with someone than to eat alone. It's often much easier to get people to "exercise"—walk, stroll, do housework or yard work, or just plain move—when they do it with someone else.

Going FAR can also be used to create effective conditions for social rest. You can walk together with a family member to wake up your cold brain in the morning, becoming more alert while obtaining morning sunlight, which enhances your mood and resets your biological clock. You can walk to lunch with a work colleague, learning about them and their family as you deepen your relationship, thin your waistline, and get mood-improving social support and sunlight. You can stroll with your family after the evening meal, connecting emotionally while talking about the day's events and perhaps stopping in to pick up a child from a play date.

7. *Going FAR can improve work performance.* Studies now find that moving after meals, especially in the afternoon, increases alertness, awareness, and the ability to focus.

Studies of exercise at lunchtime show big increases in work productivity when people move around for thirty minutes. It helps to prevent the torpor induced by our biological clocks that is commonly experienced in the early to midafternoon.

Think back to that lunchtime walk you took with a colleague, one of the many forms of social rest. Think about the potential benefits to your health, the advantages brought by movement, ranging from increased blood flow to your brain to a tighter waistline to increased alertness. You'll probably be more capable of handling whatever tasks the afternoon dishes out as your social, lunchtime walk dispels the normal midday fatigue and sleepiness. When brief naps are not possible, a midday walk with a colleague can make the rest of the workday feel easier.

8. *Going FAR makes it simpler to set times and conditions that will let you actively rest.* After you eat, you move. After you move, you rest.

9. *Going FAR may aid survival.* You now know that FAR can help you in all sorts of ways. It can also help you last long—perhaps a very long time.

It turns out the longest-lived population in the world is in the United States. They are Asian American women in metropolitan New York City.

According to Christopher Murray's Eight Americas Study, the nearly forty-nine thousand Asian American women in Bergen County, New Jersey, have an average life expectancy of 91.1 years. A much smaller group in Suffolk County has a life expectancy of 95.6 years. The daughters born in the United States are living 5 years longer than their foreign-born mothers.

How do they survive so long? They eat meals with a lot of nutritional variety. They move around quite a bit, though generally not as marathon runners or by going to gyms. Instead they walk to see friends or to the grocery store. They may garden and work around the house and yard. They are highly socially connected, and their pattern of daily behavior is regular.

Rhythm pays off. Many Asian American women in these communities practice going FAR—though they don't know that's what they're doing.

So now you know dozens of different kinds of rest techniques. All of them can be used to provide you a relaxed state of concentration and improve your overall health and performance. The patterning of FAR does far more than set conditions for a more musical day. It also positions rest as an automatic and necessary part of your life.

Tuning Your Life—Giving Yourself Permission to Rest

Rest is restoration. Without rest, our cells do not reconfigure and regrow, rebuild, and regenerate themselves. How and where they rebuild you is changed by your actions.

Remember that thought is action. Whenever we do anything, whether it's bending down to pick up a book or trying to remember a neighbor's daughter's name, different networks of brain cells fire, connections are made, and cellular communications are advanced.

All of that will leave memory traces, some conscious, many not. As we pick up that book, our legs will stress different muscle fibrils and joint cells, which will then need repair and rebuilding. Our brain will remember what the muscles felt like when they curled down to the slippery floor, while consciously we may remember only the small pinprick feeling in our stooping back. For the remainder of the day as well as during sleep, these "memories," some in the brain, many elsewhere, will percolate, shift around, be upended and transformed, creating experiences that silently embed themselves in the body's information net. All that information will then go to set the rebuilding process that constantly continues so that we may live.

It all happens through the process of rest.

In order to live well, we need to rest. We need passive forms of rest, like sleep. We need the time to rebuild our cells and organs. And we need the active rest techniques you've learned in this book because they will make us more alert and more effective. Rest more, accomplish more. Rest well, and you give yourself a better chance to live healthier and longer. So recognize that rest is a priority. And you also might want to consider for yourself the following principles:

1. *Rest is necessary to life.* We need to give ourselves permission to take the time and space required for rest.

Rest is not laziness or a useless waste of time but rather is fundamental to our health and survival. As you learn to rest effectively, you will discover many new ways to add active rest techniques to your life.

Food-activity-rest, or going FAR, represents an easy way to sequence rest in ways that help you synchronize your daily activities with your powerful internal clocks. You eat, you move, you rest. Sometimes the rest activities will be passive, like nighttime sleep or daytime naps, but since you now know how to do so many of them, you can begin to use active rest techniques more frequently. After rushing through commuter traffic and running to your workplace elevator, you can stand on that elevator in mountain pose, composing your mind and your thoughts as you rest your body. After having lunch with a colleague, you will come back to your desk more refreshed, but you can also take those few extra seconds to breathe deeply and focus on what you plan to accomplish in the next few hours, and how to complete those plans. When you're feeling gummed up and tired in the middle of the afternoon, you'll take the chance and walk to music over to a colleague's desk, asking her advice on how to solve a persnickety problem. In those few short moments you will obtain forms of mental and social rest that can revive you as well as help provide new ways to fix old, persistent dilemmas.

When you know you need rest, you can find a way to get rest. The techniques of active rest are so quick and so easy that usually several will come to your mind when you feel the need to rest.

There are other advantages to placing rest as part of the sequence of the day.

2. *Rest is musical.* If you listen to a song, you will usually hear the theme followed by the chorus, then the theme again. If you hear a piano concerto, you will hear the theme, its variations, and then the theme restated.

Rest is part of our inner biological music. Rest is like the recapitulation of a theme in that piano concerto, or the new, changed chords when we hear a song's melody a second time.

With rest, we renew. The body and brain look at what we just did, remember it, and rework and rebuild the required parts. If we are playing soccer, running down the field only to stop cold when a shot on goal goes wide, we move from activity to being very briefly at rest and then back to activity in a pattern so seamless we are barely conscious of it. But just as when we are at play we stop frequently, as when a point is over in tennis and we walk back to the baseline, there are almost always intervals between notes in our daily music. Rest can occur during many of those intervals, when our bodies and minds take stock, look at what we have done, and then take time to renew and rebuild.

One beauty of active rest techniques is that they are skills, just as playing an instrument is a skill. With time, they should become more efficient and effective. We become better players, who can move into exhilarating forms of active rest more quickly and more easily.

Food-activity-rest is one way to promote that process, as it possesses a rhythmic structure, as does playing music, play itself, and many other favorite flow activities. Food is fuel and materials, but it is also information. *When* we eat sometimes changes our bodies as much as *what* we eat. People who do not eat breakfast often have a hard time losing weight. The sequence of foods we ingest changes their overall metabolism, increasing or decreasing our risk of heart disease and cancer. Activity is what we do, but it also sets the conditions for what we will require to rebuild and retool during rest. Put together as FAR, our ordinary functions can operate like a song, with theme, variations, and restatement of the theme, playing a musical round that we can hear throughout our days and nights. Such rhythm creates sureness and stability, making it easier for our bodies to function, grow, and enjoy.

3. *Rest can be used to synchronize you with your body clocks.* Whether we are working or not, it pays to know what our body clocks are doing. We can then use techniques of active rest to realign ourselves with our inner clocks or, if necessary, to increase our alertness and concentration to better perform a task whenever we feel tired and slow.

As we now know, we emerge from sleep with a literally cold brain. It takes a while for us to wake up, to feel alert and aware. At such times active forms of rest, like taking a walk with your spouse in the early morning light (a combined social and mental rest technique) can help put us in sync with what our bodies need to do.

Similarly, during the biological downtime of early afternoon, if we are not working at a task and want to slow down, we can synchronize to that relative downtime by using potentially quieting forms of spiritual rest like moving in time and space or more meditative techniques like contemplating suchness.

However, spiritual rest techniques may or may not help us if we are feeling very tired and need to suddenly finish a job on a tight deadline. Normally our alertness levels will flow up and down throughout the day, highest in the late morning and early evening, fairly low in the early to midafternoon, lowest at night.

To rev up when we are fatigued or feeling slow, Power-Ups like self-hypnosis can revive us and alert us, allowing us in just seconds to see where we are and where we want to go while helping provide a plan to get us there. Knowing you have a tough work schedule in the afternoon can be a goad to forms of social rest like walking with a colleague to lunch, a form of social rest in which physical movement and light will help alert our bodies and brains.

Often, though, we are *too* revved up. That's when mental rest techniques like ear popping or focusing the eye can get us back in sync with where we want to be.

Around the world, many of us will feel revved up even when it's time to sleep. A sleep ritual combining several physical rest techniques like deep breathing, gravity pose, and a hot bath, or mental rest techniques like self-hypnosis, plus spiritual rest techniques like a one-minute prayer, can let us quickly reset ourselves for a deeper, more reviving sleep.

4. *Use rest techniques that you like.* You have learned more than thirty quick and easy rest techniques while reading this book. Some were easier to learn than others, some more natural to your body and personality.

Though it pays to try most of them, it makes greatest sense to use rest techniques that you really enjoy. Most people find deep breathing to be easy, often fun, and simple to use almost anywhere. As people use deep breathing, they get better at it and tend to use it again and again.

Spiritual techniques do not fit everyone. Some people have difficulty visualizing the great expanses of time and space, while a few find contemplating suchness takes considerable concentration.

Those who frequently use the social rest technique of making a special connection describe to me a sense of belonging and meaning they did not have before, plus a greater sense of common purpose.

They feel safer knowing there are people out there who will advise them when they need advice and who care about them. However, the same folks may have less use for morning family meetings. Single people, for instance, tend to like simple, quick social connects and use them often.

Self-hypnosis is loved by some people while a few find they are too nervous to become good at it (at least until they learn other rest techniques). Focusing the eye can be used by many, but others think ear popping is a bit embarrassing to do in public.

In the end, the different active rest techniques really are like music: people enjoy different kinds of music at different times of day and during different times of their lives. Fortunately, the rest techniques in this book are easy to learn and easy to remember. You can pick them up and use them even if you have not tried them for months or years.

5. *Try to use different types of rest techniques throughout the day, preferably at regular intervals.* Going FAR with food-activity-rest is one way to set a schedule. However, work, kids, relatives, and emergencies all have a way of making schedules not so regular.

Recognize that time rules life. Recognize also that having a regular daily pattern of activity and rest constitutes a large part of the lives of the longest-lived people on earth, Asian American women in metropolitan New York. Regularity of pattern is also seen among people highly effective at getting things done. To get the most out of rest techniques, it pays to use them regularly and often.

A simple rule of thumb is this: try to use at least one technique from each category—physical, mental, social, and spiritual rest—during the day. Rest techniques work best when they are conjoined and combined. Rest techniques fit quickly into an easy-to-use pattern if you are going FAR, which has the added benefit of setting your life to the rhythms of your inner biological clocks.

Of course, it can be easy to forget to rest during a busy workday. If that is the case, you might try to follow another rule of thumb: try to use a different rest technique at least once every two to three hours.

That can be hard to do if you are in the middle of a flow activity, like reading something so enthralling you don't want to stop. But our bodies are not meant to stay locked into one place for hours and hours. Humans are built to move. We are also built to rest.

One of the advantages of active rest techniques is that they don't take long to do. Within seconds you can use deep breathing or ear popping to reset yourself. More complicated techniques, like self-hypnosis, can still be accomplished in less than a minute. Many who use big screen technique on the job find themselves getting ever more efficient at it, able to get a new perspective on themselves and their work within sixty seconds.

Still, please don't set the hour chime on your watch and immediately perform a different rest technique as soon as you hear that chime—unless you really want to. Many use meals to set up times for active rest, which can take place before, during, or after you eat. Periods that in former times used to be set aside for coffee and tea breaks, like 10:30 to 11:00 a.m. and 3:00 to 4:00 p.m., are also good times to regularly do active rest techniques.

Rest techniques will be there when you need them. Even in the middle of the busiest day, you can use rest techniques to revive yourself and restore as well as review what you've been doing. Rest can help you do more with less.

6. *Rest is power.* If you can rest well, you will be sharper and quicker. Rest techniques like Power-Ups can be used to give you quick response and quick energy. Rest can reset mind and body, preparing you to ably perform tasks you might otherwise think impossible.

Rest does more than rebuild mind and body. It renews your psyche.

7. *Rest is part of the rhythm of life; use it like a dance.* Many languages use the same term for both song and dance. With luck, rest techniques can do the same for you.

Spiritual rest techniques can help us place ourselves in the world. With time and practice, you can use spiritual rest to create your own private narrative, a kind of personal life song.

But we also need to dance. The dance can occur physically, as when we walk to music. It can take place socially, as when we cook a meal with our family, going over the ingredients, what they are and where they come from, and see them transmuted into the dishes that give us sustenance and pleasure. We can mentally feel the movement of dance when we watch a tree moving in the wind, its leaves scattered by sputtering gusts.

Some of the dances we may attempt will be difficult, some of them downright hard. Some of the songs we hear in our heads are tragic and harsh, not the joyous tunes we would prefer to hear.

But even during difficult times we can still dance. Rest is active, rest is renewing, and when done with full consciousness, rest is exciting. Rest lets us become conscious of what we are and what we have.

We can use all these many rest techniques to renew ourselves, to become fully ready to do exactly what we wish to do. That's when rest truly begins to feel like its own type of dance. Most of us will perform that dance in our way and to our own purposes. Sometimes it's lots of fun to dance as if no one is looking. At those moments we may get a useful peek at our inner selves.

All you should need to make your own music in life is a little knowledge of what rest is and how to do it. You've got that now.

SELECTED
BIBLIOGRAPHY

There are lots of terrific books out there that connect issues of rest and health. Here are just a few that are useful, entertaining, or simply fun.

Barabási, Albert-László. *Linked: How Everything Is Connected to Everything Else and What It Means for Business, Science, and Everyday Life.* New York: Plume, 2002.

Carskadon, Mary. *Encyclopedia of Sleep and Dreaming.* New York: Macmillan Publishers, 1993.

Csikszentmihalyi, Mihaly. *Flow: The Psychology of Optimal Experience.* New York: Harper Perennial, 1991.

Edlund, Matthew. *The Body Clock Advantage.* New York: Circadian Press, 2003.

Gardner, Helen, ed. *The Metaphysical Poets.* New York: Penguin Press, 1960.

Gordon, James S. *Unstuck.* New York: Penguin, 2008.

Hartwell, Leland, Leroy Hood, Michael Goldberg, Ann Reynolds, Lee Silver, and Ruth Veres. *Genetics: From Genes to Genomes.* New York: McGraw-Hill, 2006.

Hauri, Peter. *No More Sleepless Nights.* New Jersey. Wiley, 1996.

Levitin, Daniel J. *This Is Your Brain on Music.* New York: Dutton, 2006.

Morton, Oliver. *Eating the Sun.* New York: Harper, 2008.

Pollan, Michael J. *In Defense of Food.* New York: Penguin Press, 2008.

———. *The Omnivore's Dilemma.* New York: Penguin Press, 2007.

Roizen, Michael, and Mehmet Oz. *YOU: On a Diet.* New York: Simon and Schuster, 2006.

Sacks, Oliver. *Musicophilia: Tales of Music and the Brain.* New York: Vintage, 2008.

———. *Uncle Tungsten: Memories of a Chemical Boyhood.* New York: Vintage, 2002.

Storr, Anthony. *Solitude.* New York: Free Press, 1988.

Wilson, Edmund O. *Consilience: The Unity of Knowledge.* New York: Vintage, 1999.

ACKNOWLEDGMENTS

Science, like our bodies, changes rapidly. I have had a lot of help from others trying to get out the newest, most useful research on rest and renewal, but many important studies will be done even in the short period between writing and publication.

Books are collaborative products, and there are many I'd like to thank, though not all can be listed here:

For conversations regarding sleep, rest, and clocks, I'd like to especially thank Mary Carskadon, Dirk-Jan Dijk, Charles Czeisler, Sharon Keenan, Sean Drummond, Charles Edwards, Gordon Stoltzner, Greg Belenky, Roseanne Armitage, David Dinges, Leon Lack, Gaby Bader, J. Terry Petrella, Larry Chilnick, and Lynne Lamberg.

For help with the manuscript, I am grateful to Carol Gaskin, Charles Edwards, Ellen Vander Noot, Laurence Tancredi, Nikitas Kavoukles, Tom Walker, Cheryl Walker, Mary LaPointe, Susan Goldberg, Joseph Mondello, Suzanne Stoltzner, Charlotte Akers, and Ian Greenhouse, among many.

With editing, I owe much thanks to my literary agent, Coleen O'Shea, who ably showed me the foibles of previous incarnations of this project; my incisive, funny editor at HarperOne, Nancy Hancock; and Priscilla Stuckey for her fine copyediting.

For help with explaining the importance of rest to the public, I am glad I've had the chance to discuss this issue with Ma Xinle, Sandy Greenhouse, Elliot Livstone, David Sensabaugh, Daphne Rosenzweig, Debbie Garofalo, Amy Lee, Michael Goldberg, Janet Steckler, Greg

Band, Anne Fisher, Irek Hicks, Chang Qing, and Julie Moberg. I also thank my office manager, Mary LaPointe, for many years of support and help.

And I owe most in many ways to my patients who lead me to new problems and push me to solve them. They explain to me what works and what does not, and keep things straight.

INDEX